Regeneration:
Key to Understanding Normal and Abnormal Growth and Development

S. MERYL ROSE
TULANE UNIVERSITY

Regeneration:
Key to Understanding Normal and
Abnormal Growth and Development

714418

APPLETON-CENTURY-CROFTS
EDUCATIONAL DIVISION
New York   MEREDITH CORPORATION

PRINTED IN THE UNITED STATES OF AMERICA

390–75871–x

Preface

In recent years the number of people studying biological problems has increased greatly. At the same time tools and methods have improved and multiplied. This has made possible many different approaches by many people to the same problem. One of the consequences of multiple approaches by specialists is a partial failure of communication between groups of specialists. There is a much greater lack of communication between the knowledge gatherers and the general public including the young who will be trying to solve the problems in the near future. In this book an attempt has been made to put a specialist's knowledge and thoughts into general terms so that they may be more widely understood.

The problem being considered is that of cellular differentiation. The emphasis is on the control of differentiation which occurs during regeneration of lost parts. Information first gained from studies on regeneration is then applied to the problems of embryonic development and tumor development.

This book was planned when the author was a Guggenheim Fellow and was written during the tenure of a Career Development Award from the Institute of Child Health and Human Development of the National Institutes of Health.

Many people have produced the work which is reported here. Special thanks are due my former graduate students and my wife for their ideas and labors. Carolyn Volpe has clarified some of the reporting with her art work. Sincere thanks to all of them.

S. M. R.

Woods Hole, Massachusetts
June 22, 1970

Contents

Introduction

Higher organisms develop in a series of steps from sperm and egg to a complex of many well-integrated parts. The sperm and egg carry all the information necessary for the production of those parts. Essentially, what Developmental Biologists are trying to learn is how different fractions of that information come to be used in different regions. What kind of control is there which leads to that well-integrated group of many different kinds of cells in the completed organism?

A complete understanding of the process requires knowledge at the various levels of organization. In recent years there have been major advances at the molecular level. Most investigators now believe that the key to understanding at this level is that different parts of the DNA complex can be turned on and off. Each differentiated region would have a different and characteristic profile of active DNA regions.

At the cellular level the question is, "What is the nature of the controlling information which travels from cell to cell and acts to turn DNA segments on or off?"

Beyond the cellular level at the organ level there are related questions. Much of the knowledge at this level comes from the study of regeneration of parts of already completed organisms. The original questions at this level were, "How does a limb stump of a salamander, for example, know that part of the limb is missing? What are the rules under which a part of that stump transforms to the missing part and nothing more?" Now the questions are becoming the same as at the cellular level. "What is the nature of the controlling information and how is it transported or transmitted?"

We shall begin by considering these and related questions at the more gross levels of organization, especially during regeneration in a variety of organisms. Motivating this exploration is the hope that general rules will emerge.

With knowledge of these general rules at the more complex levels we should be able to proceed to the less complex levels and integrate the findings in such a way that the understanding at one level may serve as the basis for understanding at adajcent levels.

Regeneration:
Key to Understanding Normal and Abnormal Growth and Development

CHAPTER ONE

Genetic Theory of Cellular Differentiation

This is the story of inquiries into how a complex organism with many parts arises from an egg—also complex but with none of those parts. It is clear that as development proceeds cells become different. For example, some will produce long processes. Others will produce hemoglobin and still others will contract because of a special arrangement of certain highly specialized proteins. One aspect of development is a series of steps within a cell leading to the special products and special function of that cell. This aspect is quite well understood and is summarized briefly in this chapter. A thorough account of the history of the problem was given by John A. Moore (1963). The major problem remaining is how development is coordinated so that each cell comes to do its special work in the proper place in a patterned whole. The continuing quest for understanding of the patterned whole is the subject of the remaining chapters.

The early knowledge of the nature of the fundamental differentiating materials came not from studies of differences within an organism but from studies of the inheritance of differences between organisms. When Gregor Mendel reasoned that a pea seed was rough or smooth because it had received certain non-blending factors, to become known as genes, from its parents, the way was open that was to lead to an understanding of the first aspect of cellular differentiation, the steps from gene to product.

An important step along the way was the demonstration that genes were parts of chromosomes. Several people had suspected that they were, but W. S. Sutton (1902, 1903) while still a graduate student, demonstrated that chromosomes pair and separate with one member of each pair going to sperm and egg in such a way that genes on them would be inherited as in Mendel's studies. Already Theodore Boveri (1902) had shown that abnormal development of sea urchins resulted from abnormal numbers of chromosomes. He had studied cell division in eggs which had been fertilized by more than one sperm. Sometimes, after two sperm had entered, three division spindles formed with the consequence that some cells received too few or too many chromosomes. These developed abnormally. Boveri reasoned that a fertilized egg to develop normally must have a complete set of chromosomes.

Also, just before Sutton's work, C. E. McClung (1901–1902) had suggested that since males have an unpaired chromosome they would produce two kinds of sperm, one determining female and the other male. This too was considered as Sutton constructed his theory of genes as parts of chromosomes.

The several inherited characters of peas which Mendel had studied behaved as if the genes for them were on separate chromosomes because they segregated independently. If genes A and B were carried by pollen or sperm and a and b by an egg the resulting offspring would be Aa Bb. The sperm and eggs for the next generation could be AB, aB, Ab or ab. The As and Bs did not have to remain together and could be inherited independently.

In the early years of this century, Thomas Hunt Morgan and his students discovered a number of inherited differences in the fruit-fly, *Drosophila*. Some moved independently as they passed from generation to generation but others were inherited in blocks, or linkage groups, as they were called. Morgan (1910) demonstrated that one group of linked genes behaved as though part of the special X chromosome. This linkage group and others apparently moved together because they were parts of the same chromosome. In time, after inheritance had been studied in a number of organisms, it could be generalized that the number of linkage groups equals the number of pairs of chromosomes. By that time it was firmly established that genes are parts of chromosomes.

Although genes could be shown to move in blocks, one or more members of a block could exchange places with part of the homologous block. Morgan and his students began to believe that this meant chromosomes paired before egg formation and that parts could be exchanged during the pairing.

A. H. Sturtevant (1913) used this hypothesis of crossing-over to determine the relative spacing of the genes. A study of many of these cross-overs indicated that the frequency of crossing-over was rather constant between two given genes, but that some pairs were separated by crossing-over more frequently than others. Sturtevant reasoned that this would be true if each gene was a special and constant spot on a chromosome, and if the distance between different genes determined the frequency that crossing over occurred between them. Imagine two parallel strings of beads wrapped around each other. If breaks came at random, the chance for a break between two beads was greater the farther apart they were. The breaks might allow recombination of the upper part of one string with the lower part of the second string. If A and C crossed over more than A and B, A and B were closer. If the sum of the frequencies of crossing-over between A and B and between B and C was roughly equal to the frequency of crossing-over between A and C then the order was ABC. Thus was the linear order of the genes established and mapping made possible.

As Professor Morgan and his students went on and made their maps of the invisible genes, many hard-nosed biologists of the day thought the whole business a foolish exercise in unreality. Then some old observations about some special giant chromosomes in flies were brought to bear on the problem. Most cells have chromosomes large enough to be seen through the light microscope only when they have shortened and thickened close to the time of cell division. The chromosomes are ordinarily so attenuated that they are invisible. In the salivary glands of flies the pairs of chromosomes of non-dividing cells are not the usual two separate strings but the two are fused and each is composed of many strings fused side by side. This makes them broad enough to be visible in the outstretched condition under the light microscope. When properly stained, bands of different thicknesses and with variable spacing between them are visible.

T. S. Painter (1934) and Calvin Bridges (1935) constructed the first salivary gland maps. One way was to choose as parents one fly with a recessive gene and another with what appeared from breeding experiments to be a total lack of genes in the same small part of the chromosome. The offspring expressed the recessive gene which they would not have been able to do if there had been a dominant gene in the same locus on the other chromosome. One could see in the paired salivary gland chromosomes exactly what part of a chromosome was missing when a gene was missing. Bands pair exactly in salivary gland chromosomes. Imagine a chromosome with genes ABCDef G. This when paired with ABCDG would express the recessives e and f. The ABCDG regions fuse leaving the portion of the one chromosome containing the e and f genes bulged out as a small loop. In such a way it was possible to map just where the genes lay. In some cases a single band with some interband space was shown to be the region within which a gene lay.

Geneticists had learned from the results of breeding experiments that occasionally the linear order of parts of linkage groups was reversed. Such inversions could be seen in salivary chromosomes. It was also known that a part of a linkage group could be transferred to another linkage group. These were visible as translocations of a part of one chromosome to another.

With the use of genetically marked chromosomes paired with abnormal chromosomes positions of genes were made visible. The maps prepared from crossing-over data and the salivary maps were remarkably alike. Genetic mapping was recognized as a major triumph of the human intellect and there were more triumphs to come.

There remained three major steps to be taken. The nature of the gene had to be determined. The pathway from gene to product and function had to be understood. Finally, we had to learn how specialized cells operating with only parts of their genome arise in certain patterned relationships.

A considerable amount of knowledge concerning what genes do was acquired before their chemical and structural nature was elucidated. During the early part of the twentieth century, it became clear that a defective gene could be transmitted from parent to offspring which resulted in the failure of a particular protein to form. Genes could somehow make proteins, but by what steps was not known (rev. by George W. Beadle, 1951).

George Beadle and Boris Ephrussi (1937) worked out a method for studying the functional relationships of genes. By the time they started their work, there were many different eye colors known in the fruit-fly, *Drosophila*. What they did was to transplant discs on their way to becoming eyes from one genetic type to another. This was done before the eye pigments had formed. It seemed possible that products of gene action might diffuse from host to graft and affect the color of the eyes. The common wild type eye color is a dark red. Their genetic stocks each had a certain specific eye color with names such as vermillion, claret and cinnabar. Certain mutant eye discs could become dark red wild type eyes when grown in genetically wild type hosts. It seemed that the mutant grafts could use some diffusible product from the host which the grafts could not make. Furthermore, by transplanting eye discs to various hosts differing in eye color genes, results were obtained which indicated that the production of eye pigment is dependent upon a series of genes. In one series, a genetically light red grown in a medium red host received something from the host which enabled the graft eyes to make it all of the way to dark red. The medium red hosts lacked something enabling them to become dark red. The grafts lacked something enabling them to change light red to medium red. The genetically light reds received what they lacked from the medium red hosts and could then go all of the way to dark red. The grafts were defective in that they could not convert light red to medium red but once that had happened they could use the last gene in the series which converted medium red to dark red. The hosts lacking that gene remained medium red.

When a genetically medium red disc was grown in a genetically light red host, the host, blocked in the first step—light red to medium red— was not making anything which would change medium red to dark red and had nothing to give the graft. It remained medium red.

The interpretation was that there were two steps. One was from light red to medium red and the second from medium red to dark red. Because biological reactions are in general enzymatically catalyzed, it was reasoned that each gene was responsible for a different enzymatic protein. First one acted and then the other. This was the beginning of our understanding that each of the many biological reactions yielding products for other reactions is controlled by a different gene.

To pursue the relationship of genes and enzymes further, Beadle collaborated with the biochemist, Edward Tatum (1941, 1945). A very clever experiment was planned. This time, instead of working with naturally occurring mutants, they produced new ones. H. J. Muller (1927, 1950) had previously demonstrated that X-rays damage spots on chromosomes and can knock out genetic activity of those regions. Beadle and Tatum chose the bread mold, *Neurospora*. Their plan was to irradiate and then test identical spores for their ability to produce growing mycelia on two different media. The one medium called minimal contained only agar, water, salts, a carbon source and biotin. Normal *Neurospora* can grow on this medium. The other medium was supplemented with amino acids, vitamins and various other substances which might be important to growth. The rationale was that any induced mutant which could not grow on the minimal medium but could grow on the supplemented medium had lost the ability to make something in the supplement. By testing with the various ingredients it could be determined what the mutant needed.

Sometimes more than one gene defect could be relieved by the addition of one substance. For example, tryptophan caused several different mutants to grow. One of the mutants needed tryptophan and only tryptophan.

Another, if supplied with indole, could make its own tryptophan. Still another could grow if given any one of the three compounds, tryptophan, indole, or anthranilic acid. If given anthranilic acid it could make indole and, from indole, tryptophan could be produced. Tryptophan is an amino acid used in the construction of proteins. Tatum and Beadle learned in several instances that the chains of reactions already discovered by biochemists in mammalian liver were the same in *Neurospora*. Two important things were being learned. Biochemistry is essentially the same in molds and mice, and each of the reaction steps is controlled by one gene.

Much was being learned about gene controlled activity, but the nature of the gene was still unknown. Most investigators were guessing that the genes were proteins, but this was to turn out to be wrong. The first observation which was to lead to a demonstration of the nature of the hereditary material was an unexpected one. Fred Griffith (1928) was trying to develop cures for various kinds of pneumonia caused by bacteria. Pneumococci of different strains were known which differed by the kind of polysaccharide making a capsule around the bacterium. Sometimes a pneumococcus can lose its special polysaccharide-determining gene and then it cannot be typed for its polysaccharide. The descendants of capsule-free pneumococci are also capsule-free and are less virulent. At the time Griffith was injecting two kinds into mice. One was a virulent Type III with capsules, and the others had descended from Type I, but one of their ancestors had lost its special capsule-forming gene. These capsule-free bacteria were not virulent. In the hope that a useful vaccine might be produced, Griffith injected living, non-virulent, typeless, and heat-inactivated Type III into the same mice. The heat was rather mild, about 55°C, and prevented further growth of the Type IIIs when they were injected alone. The surprising thing was that when the so-called heat-killed bacteria were injected along with the typeless bacteria, the mice became very sick. Alone neither would have had much effect. The pneumococci from the sickened mice were examined. They turned out to be Type III. The only reasonable explanation seemed to be that something from the killed, but not chemically mutilated, Type IIIs had been added to the former Type Is, then typeless, converting them to Type III.

O. T. Avery, Colin Macleod and Maclyn McCarty (1944) systematically searched for the transforming agent. Removal of the capsule and much of the protein still left a potent transforming material. Extracts of deoxyribose nucleic acid (DNA) turned out to be the hereditary material. It could be extracted from cells of one type and cause other cells to change *in vitro* to the donor type and make more of the donor type DNA. The DNA was not like X-rays and chemicals which caused mutations. These mutagens made random hits and could affect any gene. The DNA which was being transferred acted differently. It was itself the genetic material. Once in the new host, it became part of the host and could, with the aid of the host, replicate and in addition cause the production of a particular polysaccharide.

Other work indicated that DNA was the substance of genes. Chromosomes had been known to be part DNA and part protein, and later it was to be known that another kind of nucleic acid, ribose nucleic acid, or RNA, was also produced in chromosomes. The works of A. E. Mirsky and Arthur Pollister (1943) and Hewson Swift (1950) made it clear that all normal cells of a species contain the same amount of DNA, but that the other components of chromosomes vary in amount. Work to be considered later indicated that all cells of an organism contain all of the genetic information. The constancy of the DNA from cell to cell had made it the likely bearer of the genetic information, the same in every cell.

Then A. D. Hershey and Martha Chase (1952) demonstrated that when a bacterial virus imparts genetic information, it is only the nucleic acid which enters the bacterium. The protein coat of the virus is left outside. Furthermore André Lwoff (1953) was able to present convincing evidence that viral nucleic acid could join with normal nucleic acid in bacteria and become part of the inherited DNA of the host. Added proof was given by Joshua Lederberg (1960) and Norton Zinder (1958), who demonstrated that viral genetic material could combine with bacterial genes and carry those normal genes to new host bacteria. There was no doubt by this time that DNA was genetic material. It was also known that some viruses contain only RNA. It too could carry genetic information.

Another important fact became apparent during the forties and fifties. Torbjoern Casperson (1950) and Jean Brachet (rev. 1957) were able to demonstrate that where there was much synthesis of protein, much RNA was present. The synthesis of protein appeared to occur on small cytoplasmic particles now known as ribosomes or polysomes when they are in groups. It seemed that RNA had to be involved in the production of proteins. It was also clear that DNA was involved. There had to be some relationship between DNA and RNA. Their ingredients are similar. Evidence was collecting that RNA was made in the nucleus where the DNA is, but that the RNA migrated to the cytoplasm where most of it was to be found and where it participated in the production of proteins. The idea was growing that RNA is the genetic deputy of DNA.

There seemed to be a peculiar urgency among biologists to learn more about these nucleic acids. Much was learned in a few short years. Avery and his coworkers had established that DNA is hereditary material in 1944. By 1953 James Watson and F. H. C. Crick took the pertinent information which had accumulated and used it to construct a model of a DNA molecule which has stood the test of time. They used several kinds of information. They knew the ingredients. There was a five-carbon sugar known as deoxyribose. There were four different nitrogenous bases, adenine, guanine, thymine, and cytosine, and also phosphoric acid. All ingredients were repeated many times in a molecule. Watson and Crick were looking for something which could be a model for more of itself. Genetic knowledge required this. It also had to carry information for the making of particular proteins. Before the task of constructing a reasonable model was done, such a task seemed beyond man's intellect. When it was done, it was beautifully simple and today is understood by many people.

Besides the ingredients and the fact that they repeated, Watson and Crick knew from crystallographic analyses that the molecule probably consisted of two chains. By 1953 M. F. H. Wilkins, Rosalind Franklin, and their associates had worked out in DNA the spacing of atoms and their angles with respect to each other. (Franklin and Goslin, 1953; Wilkins, 1956.) When Watson and Crick alternated molecular models of the sugar and phosphate with the proper angles to each other, they had a helix. When two helices were put together and connected by the nitrogenous bases in a particular way, a spiral staircase had been formed with pairs of bases as the steps.

The clue as to how the steps should be constructed came from studies begun by Erwin Chargaff (1951, 1955). He and his associates had analyzed DNA from several microorganisms and had learned that the ratio of the purines, adenine and guanine, to the pyrimidines, thymine and cytosine, was 1.01. Figuring that the .01 represented error, this would mean that every time a purine was used in DNA there had to be a pyrimidine. The ratios of adenine to thymine and guanine to cytosine were also close to 1.0. G. R. Wyatt (1952) took the analysis a step further, and learned that the amount of adenine was always the same as the amount of thymine in 11 different insect viruses. Furthermore, guanine and cytosine were equal in all 11. The totals of $A + T$ and $G + C$ varied considerably between the viruses.

Watson and Crick took the above information to mean that A and T were always paired and so were G and C. In the model, only when an A attached to the sugar of one spiral was opposite a T of the other, and when G and C were similarly paired, did the double helix curve smoothly without bulging or sagging at spots.

Almost immediately it was understood (Watson and Crick, 1953b) how more DNA could be made identical to itself. If the two coils separated and each A attracted a T, each T an A, each G a C and each C a G, and if each base being attracted came with an attached deoxyribose, and a phosphate group was added between sugars, each of the two original coils would have attracted, in the proper order, the ingredients of the complementary sister coil from which it had recently separated. If the original pair were A and B, a new B would have been made on A and a new A on B. Experiments done to partially test this idea yielded the result that the old strands remain intact while new ones are made on them. Chromosomes or DNA were labeled with radioactive isotopes during a first replication and then allowed to replicate further in the presence of non-radioactive substrates. The originally labeled strands retained the label and remained intact, except in cases of crossing-over, and new unlabeled strands arose in combination with them (J. Herbert Taylor and coworkers, 1957 and M. Meselson and F. W. Stahl, 1958).

The model not only suggested how replication could occur but also how RNA could be formed in a similar manner. The ingredients of RNA are slightly different from those of DNA. Whereas the deoxyribose of DNA has a hydrogen attached at one spot, the similar ribose of RNA differs only by having an hydroxyl group at that spot. The sugars in both are linked by phosphate groups. The major difference in the ingredients of the two types of nucleic acids is that in RNA uracil instead of thymine is used as a code letter. Uracil differs from thymine by having a hydrogen atom in place of a methyl group. It was suggested that when RNA was being made, each base of the DNA attracted its complementary base, but that adenine attracted uracil rather than thymine.

R. H. Doi and Sol Spiegelman (1962) showed that the order of bases in DNA determines the order of bases in RNA. DNA and RNA were extracted from an organism. The DNA was labeled with one isotope and the RNA with another. When the DNA and RNA were subsequently mixed the two labels came together, indicating that the RNA which had been transcribed from a particular DNA was so much like it that they could pair in complementary fashion. DNA and RNA from unlike organisms do not pair or hybridize.

The realization that there were orderly sequences of bases in the RNA led to a consideration of the possibility that the sequences of amino acids in proteins might be determined by the base sequences in the RNA. There was hope that the "language" of the nucleic acids could be deciphered. The next year after the year of the model, George Gamow (1954) reasoned that since there are 20 primary amino acids used in proteins, the least number of bases in sequence needed to code for each was three. To use the now common analogy, one must write three-letter words to make 20 different words from a four-letter alphabet (ATGC). Actually, 64 three-letter words can be written, One cannot write the necessary 20 words using only two-letter words.

Crick (1962) went on to show that indeed the code is read in groups of three. He studied mutants of bacterial viruses which apparently differed from normal by lacking a single base in their sequence of bases. These could not code for a functional virus protein. Crick's idea was that the translation of the base sequence into an amino acid sequence would fail if a single letter was missing. As an example, suppose the sequence of bases at the beginning of a stretch of DNA were GGCTAATTGCCA. Its complementary sequence in RNA would be CCGAUU(A)ACGGU. If the circled A were lost and the message was translated in groups of three from left to right, CCG could mean an amino acid, AUU would mean another, and the first part of the protein would be normal. Because of the defect, the rest of the message would be translated improperly. Instead of AAC being translated, it would be ACG and then GU. The translation of three letter words would be out of phase beyond the defect. Beyond the single lost base, the next group of three would be the letters for the last two-thirds of a word and the first letter of the next word. The translation would be out of phase from there on and the protein quite different from normal. By combining three different but closely spaced single minus mutations, the message after the last mutation in the series would be translated properly. Crick found that combinations of plus and minus mutations which reframed the translation so that the original message and the translation were in phase with a frequency of three did give functional proteins. These of course would have had short abnormal portions, but not all parts of large protein molecules are vital to their function.

In the meantime, a great deal had been learned about the enzyme systems, the polymerase systems, used in the assembly of DNA and RNA. Severo Ochoa (rev. 1963) and his associates had gone so far as to produce RNA *in vitro* and Arthur Kornberg (rev. 1959) had learned how to produce DNA *in vitro*. Everything was ready for an attempt to learn how to translate base sequences of RNA into amino acid sequences in proteins. Marshall W. Nirenberg and J. H. Matthaei (1961), working at the National Institutes of Health in Bethesda, Maryland, performed the following experiment. An *in-vitro* protein-producing system of enzymes and substrates was assembled. All 20 amino acids were added, but instead of adding a natural RNA with its code for a particular protein, they added a polymer of just one of the bases, uracil. This poly U message was read, and a protein was produced. It consisted of only one amino acid, phenylalanine, repeated many times. It turned out that UUU meant phenylalanine. Again, what had been speculation based on genetic studies had been confirmed.

In a short time Nirenberg and his associates and Ochoa and his associates had learned which amino acid each of the base triplets coded for. Our example on the preceding page, GGCTAA in DNA is transcribed as CCGAUU in RNA and is translated as arginine-isoleucine in a protein.

The model has served as an invaluable hypothesis. Even mutation has been accounted for. It was already known from the studies of Linus Pauling and his associates (1949) that the inherited abnormality known as sickle-cell anemia results from the production of abnormal hemoglobin. Vernon Ingram, after breaking up normal hemoglobin and hemoglobin from sickled cells into fragments, compared the fragments. It turned out that the substitution of a single amino acid in normal hemoglobin is responsible for this kind of anemia. Function and form of red blood cells are changed by replacement of a single amino acid. It now appears that a change in only one base in DNA may lead to the substitution of an amino acid. It was also known from genetic studies on a virus by Seymour Benzer (1955) that a cross-over may occur in the space occupied by two DNA bases and presumably between them. Now Charles Yanofsky (1963) has shown that an amino acid substitution, a genetic mutation, may result from a substitution of a single base (or letter) in the nucleic acid code, changing one three-letter word to another.

Many scientists have contributed to the work leading from Mendel to Molecular Biology. Important steps were taken by the people mentioned above. There are many more not mentioned here (*see* John Reiner, 1968), because most of our space is devoted to a study of the system in which different parts of the code come to be translated in different parts of an organism. Together, these scientists had brought us much closer to an understanding of the chemistry and basic structure of life. The human intellect had had its greatest triumph.

The next problem is that of differentiation. How can a certain part of the DNA be used in some cells and another part in different cells? The suggestion of how this might occur in multicellular organisms comes from other studies on unicellular organisms (J. Monod and F. Jacob, 1961; F. Jacob and J. Monod, 1961). They were working with the genetic control of enzyme production. The particular enzyme was one which splits the sugar, lactose, and other β-galactosides. The enzyme is known as β-galactosidase. It is not present in the cells at all times. It appears only when a β-galactoside substrate is in the medium. When glucose is the sugar being used and no β-galactosides are present, no β-galactosidase is produced. The enzyme is known as an inducible enzyme. In this case, the substrate induces the enzyme which will use the substrate. In contrast, many enzymes are present at all times. These are the constitutive enzymes.

Several mutations affecting the metabolism of β-galactosides were detected and studied. One of the mutations changed the system so that the enzyme continued to be produced after the substrate was removed. An inducible enzyme had become a constitutive enzyme. Other mutations were discovered which resulted in the production of abnormal enzyme molecules, like the enzyme in some ways, but with little or no enzymatic activity.

The system which Jacob and Monod suggested as the one best able to encompass all of the facts was one in which a number of genes called *structural genes* contained the ultimate information for the enzyme molecule. The structural genes were recognized as multiple because there were different mutations affecting structure. This as we now know might be a continuous stretch of DNA. At the same time, there had to be a genetic locus where transcription could be started or repressed. This they called an *operator gene*, and operator plus the adjacent structural genes were known as an *operon*. When the operon was functioning normally, the normal enzyme was produced.

To account for the fact of inducibility and that the inducible system could become constitutive, a regulator gene was inserted. This need not be immediately adjacent to the operon. Its function was to produce a repressor which would block transcription from starting at the operator. When glucose was the sugar being used, the regulator gene produced a repressor which would inactivate the galactosidase operator gene, and no galactosidase was produced. When lactose was added, the lactose was thought to combine with the repressor substance, thus preventing it from blocking the operator gene (A. B. Pardee, F. Jacob and J. Monod, 1959). In this condition, with the operator not blocked, the structural genes could make their RNAs which would be translated to β-galactosidase and other enzymes coded for in the operon. A gene was on when not repressed. Induction of the activity of a particular gene required its derepression.

When the regulator gene mutated and could no longer produce the specific repressor, the operator which it normally controlled was left in the permanently on condition. This accounted for the change from inducible to constitutive.

Jacob and Monod suggested that differentiation in multicellular organisms could result from the passage of metabolites between the cells which would derepress. Thus a product of one cell might induce an enzyme in another.

Other kinds of studies given throughout the book indicate that repression of some genes already on may be the major way in which differentiation is achieved. The theory of differentiation by inhibition was published independently by four investigators in the years 1952 to 1954. Many studies with embryos and regenerates indicate that early in differentiation more genes are on in the cells than will be when they achieve their final specialization. Differentiation involves a gradual change from many possibilities to one actuality. The basic idea presented in the early fifties was that a region first to differentiate could produce specific repressors or inhibitors, which would block the same gene-based reactions in adjacent regions to which it was transported. The cells of these regions would then employ other parts of the genome. They in turn would prevent their neighbors downstream from achieving the second-level genetic status. Thus each region would differentiate as it received repressive messages from upstream (S. M. Rose, 1952).

H. V. Brøndsted (1954, 1955) one of the contributors to the theory of differentiation by repression, had done considerable work on regeneration in planarian worms. If one cuts a worm into several parts, one learns that all or most parts can form a head but that they do this at different rates. Charles Manning Child (1941) had shown that this is true in all regenerating organisms. There is an orderly decrease in the rate of head formation the farther away from the head one takes the isolate. The most thorough study had been one on *Tubularia*, a marine hydroid, by Lester G. Barth (1938). What impressed both Barth and Brøndsted was that if parts were combined, it was always the one with the highest rate of head formation that actually formed a head. It somehow inhibited head formation in its slower neighbors when they were combined.

This was also shown to be true for parts of heads. When a head is amputated, cells in the most anterior part of the remainder transform to the missing part. If while this was happening Brøndsted cut the young regenerate into medial and lateral pieces, both formed eyes. The medial portion produced eyes more rapidly than lateral pieces. When left together, only medial portion produced eyes. As we shall see, this is a universal phenomenon. The fastest regions produce certain structures and at the same time prevent their slower neighbors from producing those structures. This means that the dominated regions are reduced to using another part of their genome. More about the specificity and action of the repressors is given in the following chapters.

Martin Lüscher (1953), another who contributed to the theory of differentiation by repression, used two bodies of knowledge. He knew a great deal about regeneration, including the fact that as long as a particular structure is present it prevents like structures from forming. Lüscher also studied the same kind of relationship during differentiation of the castes in termites. Termite nymphs about to transform to reproductives do not do so if older reproductives are already present. These studies of caste differentiation in termites had been initiated by S. F. Light (Light and Weesner, 1951). Lüscher went on to show that sexually differentiated forms need not be in the same cage to suppress sexual development in younger termites. However, the cages must be close enough so that the older already differentiated sexual forms can touch antennae of some members of the group being inhibited. It is believed that a hormone is passed from the already differentiated which suppresses like differentiation in those still differentiating. The social hormone emanating from queens and limiting the production of queen bees and queen ants has also been demonstrated (C. G. Butler, 1954).

Werner Braun (1952), another of the four who proposed that repression might be the mechanism of differentiation, worked with bacteria. His cultures of *Brucella abortis* demonstrated the common growth pattern. A small inoculum grew slowly at first, then rapidly reached a plateau and finally the number of bacteria decreased. Braun noted that the amino acid alanine was an excretion product. When growth stopped, a certain concentration of alanine had been produced. That same concentration of alanine added to rapidly growing cultures stopped the growth of the bacteria. Here was a case in which a product of growth was limiting growth.

More important to Braun was the fact that as his original type of *Brucella* began to decline, some always transformed to a particular genetic variant. The second type in its turn grew and declined. The variant produced alanine more slowly and was stopped by it at a higher concentration. As the second type stopped growing still another type appeared and completed its growth cycle. Braun realized that here was a differentiating system and that such a serially self-limiting system of multiple possibilities might be the basis of differentiation in multicellular organisms.

A paper which I had written and Braun's paper appeared together in the *American Naturalist* in 1952. It was suggested in mine that only a hierarchy of genes with a system of orderly serial inhibition of them could account for many of the facts known about embryonic development and regeneration. If gene A were in use in the most rapidly developing cells, a product of A would pass to nearby more slowly developing cells and repress their A gene. These would then use B. When B products reached a sufficient concentration, B genes in still slower cells would be blocked. They would then use gene C. So it would go until all genes in the hierarchy were in use. Braun, using an entirely different set of facts, suggested essentially the same thing.

Also in 1952, beginning with the work of H. Vogel and B. D. Davis (rev. Vogel, 1957), it became clear that quite generally end-products as they accumulate can prevent the formation of an enzyme in the chain leading to the formation of the product. In the original case it was demonstrated that arginine added to a bacterial culture could prevent the formation of acetylornithinase, one of the enzymes along the pathway leading to arginine. Later it was learned that a variety of substances including other amino acids, purines and pyrimidines can act in a negative feedback system to regulate their own production (H. E. Umbarger, 1963). Possibly such small molecules could act in cellular differentiation, but it now appears that these regulate production rather than turn off genes more or less permanently as occurs in cellular differentiation.

Evidence is increasing which indicates that repressors may act at the DNA level or at the RNA level, and that they are proteins. B. R. McAuslan (1963) added pox virus to cultures of HeLa cells, a strain of human tumor cells. Shortly after the viral genes started to function, viral thymidine kinase (TK) appeared. Ordinarily the TK is produced for 6 hours and then its formation is stopped. If the antibiotic, Actinomycin D, which interferes with the production of the coded messenger RNA (mRNA) was added at different times during the first two hours, the earlier it was added the less TK was formed. Actinomycin was blocking the transcription of the mRNA which translated as TK. Treatments with Actinomycin D started more than two hours after infection did not affect the production of the TK mRNA. It had all been made. On the contrary, the antibiotic when added between 2 and 4 hours after infection increased the time of action of thymidine kinase from the normal 6 hours to 17 or more hours. This means that another mRNA involved in repressing the TK system is made in hours two to four. Apparently the repressor is a protein, because it is not produced in the presence of Puromycin, which interferes with translation from mRNA to protein. This and some other repressors may act, not at the DNA level, but on the mRNA of the polysomes (A. S. Spirin, 1966).

Later, after eggs have cleaved into many cells, the differentiation which occurs seems to involve which masked mRNAs will be used in which cells (A. Tyler, 1967). The RNAs produced at a given time in a developing egg may not be used until a day or two later. The evidence for this is the delay of a day or two in the production of multiple abnormalities after a short treatment with Actinomycin D (Jack Collier, 1966). It is also clear that more DNA genes are producing more different masked mRNAs per cell during cleavage than their descendants will be producing later. Differentiation also involves shutting off transcription of much of the DNA as the cells become specialists. Individual cells may be limited finally to the production of only one or a few special proteins.

The question of how a particular gene can be turned off is being asked repeatedly. The Stedmans (1950) suggested that the very basic proteins known as histones and found in abundance in chromosomes could complex with parts of the DNA and render it inactive. That this may indeed be the way genes are turned off is suggested by a series of experiments. The DNA of thymus nuclei minus histones helps produce protein more rapidly than when histone is present (V. G. Allfrey, 1961; V. G. Allfrey, V. C. Littau and A. E. Mirsky, 1963). The same is true for DNA from pea seedlings (Ru-Chih Huang and James Bonner, 1962, rev. Bonner, 1965).

There seem not to be enough different histones to repress the multitude of different DNA coding loci. The natural repressors appear to be complexes with RNA (P. S. Sypherd and N. Strauss, 1963). Complexes of RNA and histone are found in pea seedlings (R. C. Huang and J. Bonner, 1965) and in mammalian cells (W. Benjamin et al., 1966). The suggestion is that the RNA would give the necessary specificity. It would be able to complex with a stretch of DNA complementary in base arrangement to itself.

Differentiation might begin as an RNA-histone was released when its homologous DNA region became active. If it were subsequently transported and bound to the homologous DNA of another cell, the second cell would be less capable of using that particular stretch of DNA. Throughout the book many cases of differentiation are presented which indicate that differentiation occurs as one region prevents another from duplicating its activity.

The first demonstration that histones (probably RNA-histones) are involved in differentiation was given by James Bonner, Ru-Chih Huang and R. Gilden (1963). They started with buds and cotyledons of germinating pea seeds. The cotyledons produce a seed-globulin which is not found in buds. The chromosomes from buds do not produce seed-globulin in a cell-free system capable of producing proteins. However, when DNA from buds was stripped of its histones it did code for a small amount of seed-globulin.

On the other hand, chromosomes from cotyledons help produce a much larger proportion of their proteins in the form of seed-globulin. When the histones are stripped from chromosomes of cotyledons leaving just DNA, the percentage of protein which is seed-globulin is much less —in fact, the same percentage as is produced under the action of stripped DNA from buds.

It is clear that the chromosomes of buds and cotyledons are different in that those from cotyledons can produce much seed-globulin. Those from buds can produce none. When the histones are removed from both, the DNA from both is alike and can supposedly support the production of many proteins, a small fraction being seed-globulin. This demonstration, plus the knowledge that different portions of the DNA code for different proteins, leads to the belief that the pattern of DNA repression by histones is the basis of differentiation.

After the possible role of RNA-histone in differentiation in higher plants was elucidated, studies with the bacterium, *Escherichia coli*, revealed protein repressors which are not histones. It has now been clearly established that the *lac* repressor and the repressor for lambda phage are acidic proteins. Both function by attaching to the operator gene, thereby preventing transcription of specific stretches of DNA (rev. by Mark Ptashne and Walter Gilbert, 1970). It seems too early to know with certainty whether either or both kinds of repressors are the agents of cellular differentiation.

It appears from work cited above, and from other kinds of studies described in later chapters, that when cells act in concert to produce a variety of types, the major agents of control are repressors. It is also known that there are hormones which can activate or derepress inactive portions of the DNA and thus produce new differences. Changes in levels of hormones trigger the massive genetic changes which occur at metamorphosis. When, for example, a fish-like tadpole transforms to a frog, or when a worm-like insect larva metamorphoses to a fly or a moth, hormones are the initiators of the change. Concurrent changes can be observed in the giant chromosomes of larvae of Diptera. As we shall see, there is a causal relationship between changes in hormone level and changes in the giant chromosomes.

Before approaching the experiments with hormones, we must introduce the proof at the visual level that differentiation involves the use of different portions of the DNA (rev. by Joseph Gall, 1963). W. Beermann, beginning in 1952 and followed by others (rev. Beerman, 1959), demonstrated that some regions of dipteran giant chromosomes are enlarged into puffs. The chromosomes of each organ have patterns of puffing specific to that organ. Orderly changes are observed in each pattern during the course of development. The puffs of these chromosomes and the similar loops of the lampbrush chromosomes in actively growing oocytes are sites of rapid RNA production. They incorporate tritiated uridine, indicating that they are making RNA, whereas unenlarged chromosomal regions are not making it (J. G. Gall and H. G. Callan, 1962). A puff or a loop is an indication of a functioning gene. At any one time there are relatively few genes turned on in each cell.

It was possible for Beermann (1961) to demonstrate that a certain puff on a salivary chromosome was responsible for the production of granules in saliva. He worked with two species of midges, *Chironomus tentans* and *C. pallidivittatus*. Only in the saliva of the latter species are granules present, and they are produced by only four cells of the salivary gland. Chromosome IV has a puff at the same band in the four granule producing cells. The same region of Chromosome IV is not puffed in the other cells of the salivary gland of this species. The similar chromosomal region is also not puffed in the corresponding but non-granule producing cells of the other species.

The two species can be crossed. The chromosomes of the hybrid offspring form pairs with one member coming from the father and one from the mother. In the proper place in Chromosome IV of the four cells which produce granules, one half is puffed and the other half is not. This was proof that puffing indicates a gene in action.

Now we return to the demonstration that a hormone can turn on a gene. U. Clever (1961), also working with *Chironomus*, learned that as the time for pupation approaches, a series of new salivary gland chromosome puffs arise in a certain order. When the insect hormone, ecdysone, was injected into larvae not yet ready to pupate, premature pupation was induced. The first in the same series of puffs was also induced within minutes after injection of the hormone.

It is quite clear now that hormones can directly or indirectly turn on certain genes all over an organism at the time of metamorphosis. It is a means for coordinating a change all over the body from larval to adult organs. It is used in addition to differentiation by repression.

The rest of the book deals with the manner in which cells communicate with each other, with the result being a complicated, integrated organism. In most of the book the experiments discussed were performed on regenerating organisms rather than on eggs. The main advantage is that one can study how one or several parts arise, rather than start with the more difficult task of determining how the complexity of an egg becomes the very different complexity of an embryo. Once principles of differentiation were learned from regeneration studies, it was possible to use them in conjunction with studies of embryonic differentiation to more fully understand embryonic development.

References Cited

Allfrey, V. G. 1961. Observations on the mechanism and control of protein synthesis in the cell nucleus. Reprint 140, *Symp. II. Proc. Fifth Intern. Congr. Biochem.* Moscow. In O. Lindberg, ed. *Functional Biochemistry of Cell Structure*, Vol. 2:127–147. Oxford: Pergamon.

Allfrey, V. G., Littau, V. C., and Mirsky, A. E. 1963. On the role of histones in regulating ribonucleic acid synthesis in the cell nucleus. *Proc. Natl. Acad. Sci.* U.S. 49:414–421.

Avery, O. T., Macleod, C. M., and McCarty, M. 1944. Studies on the chemical nature of the substance inducing transformation of pneumococcal types. Induction of transformation by a desoxyribonucleic acid fraction isolated from pneumococcus type III. *J. Exp. Med.* 79:137–158.

Barth, L. G. 1938a. Quantitative studies of the factors governing the rate of regeneration. *Biol. Bull.* Woods Hole 74:155–177.

———. 1938b. Oxygen as a controlling factor in the regeneration of *Tubularia. Physiol. Zool.* 11:179–186.

Beadle, G. W. 1951. Chemical genetics. In *Genetics in the 20th century*. ed. L. C. Dunn, pp. 221–239. New York: Macmillan.

Beadle, G. W. and Ephrussi, B. 1937. Development of eye colors in Drosophila: diffusible substances and their interrelations. *Genetics* 22:76–86.

Beadle, G. W. and Tatum, E. L. 1941. Genetic control of biochemical reactions in *Neurospora. Proc. Natl. Acad. Sci.* U.S. 27:499–506.

———. 1945. Neurospora II. Methods of producing and detecting mutations concerned with nutritional requirements. *Amer. J. Bot.* 32:678–686.

Beermann, W. 1959. Chromosomal differentiation in insects. In *Developmental Cytology*, ed. D. Rudnick, pp. 83–103. New York: Ronald.

————. 1961. Ein Balbiani-Ring als Locus einer Speicheldrusenmutation. *Chromosoma* 12:1–25.

Benjamin, W., Levander, O. A., Gellhorn, A., and DeBellis, R. H. 1966. An RNA-histone complex in mammalian cells. The isolation and characterization of a new RNA species. *Proc. Natl. Acad. Sc* . U.S. 55:858–865.

Benzer, S. 1955. Fine structure of the genetic region in bacteriophage. *Proc. Natl. Acad. Sc* . U.S. 41:344–354.

Bonner, J. 1965. *The molecular biology of development.* 155 pp. Oxford and New York: Oxford University Press.

Bonner, J., Huang, Ru-Chih, and Gilden, R. 1963. Chromosomally directed protein synthesis. *Proc. Natl. Acad. Sc* . U.S. 50:893–900.

Boveri, Th. 1902. Uber mehrpolige Mitosen als Mittel zur Analyse des Zellkerns. *Verh. der Phys.-Med. Ges.* Würzburg 35:67–90.

Brachet, J. 1957. *Biochemical Cytology.* pp. 516. New York: Academic Press.

Braun, W. 1952. Studies on population changes in bacteria and their relation to some general biological problems. *Amer. Nat.* 86:355–371.

Bridges, C. B. 1935. Salivary chromosome maps. *Jour. Hered.* 26:60–64.

Brøndsted, H. V. 1954. The time-graded regeneration field in planarians and some of its cytophysiological implications. *Proc. 7th Symp. Colston Res. Soc.,* Colston Papers 7:121–138.

————. 1955. Planarian regeneration. *Biol. Revs. Cambridge Phil. Soc.* 30:65–126.

Butler, C. G. 1954. The method and importance of the recognition by a colony of honeybees (*A. Mellifera*) of the presence of its queen. *Trans. R. Ent. Soc.* London 105:11–29.

Casperson, T. 1950. *Cell Growth and Cell Function.* pp. 185. New York: W. W. Norton.

Chargaff, E. 1951. Some recent studies on the composition and structure of nucleic acids. *Jour. Cell. Comp. Physiol.* 38: Suppl., 41–59.

————. 1955. Isolation and composition of the deoxypentose nucleic acids and of the corresponding nucleoproteins. *In The Nucleic Acids 1,* ed. Chargaff and Davidson, Chapt. 10, pp. 307–372. New York: Academic Press., Inc.

Child, C. M. 1941. *Patterns and Problems of Development,* pp. 811. Chicago, Illinois: University of Chicago Press.

Clever, U. 1961. Genaktivitäten in den Riesenchromosomen von *Chironomus tentans* und ihre Beziehungen zur Entwicklung. *I. Genaktivierungen durch Ecdyson. Chromosoma* 12:607–675.

Collier, J. 1966. The transcription of genetic information in the spiralian embryo, in *Current Topics in Developmental Biology,* ed. A. Monroy and A. A. ʹMoscona. Chapt. II, pp. 39–59. New York: Academic Press.

Crick, F. H. C. 1962. The genetic code. *Sci. Amer.* 207(4):66–74.

Doi, R. H. and Spiegelman, S. 1962. Specific hybridization test. *Science* 138:1270.

Franklin, R. E. and Goslin, R. G. 1953. Molecular configuration in sodium thymonucleate. *Nature* 171:740–741.

Gall, J. G. 1963. Chromosomes and cytodifferentiation in *Cytodifferentiation and Macromolecular Synthesis*, ed. M. Locke, pp. 119–143. New York: Academic Press.

Gall, J. and Callan, H. G. 1962. H³-Uridine incorporation in lampbrush chromosomes. *Proc. Natl. Acad. Sci. U.S.* 48:562–570.

Gamow, G. 1954. Possible relation between deoxyribonucleic acid and protein structures. *Nature* 173:318.

Griffith, F. 1928. The significance of pneumococcal types. *J. Hyg.* 27:113–159.

Hershey, A. D. and Chase, M. 1952. Independent functions of viral protein and nucleic acid growth of bacteriophage. *Jour. Gen. Physiol.* 36:39–56.

Huang, R. C. and Bonner, J. 1962. Histone, a suppressor of chromosomal RNA synthesis. *Proc. Natl. Acad. Sci. U.S.* 48:1216–1222.

———. 1965. Histone-bound RNA, a component of native nucleohistone. *Proc. Natl. Acad. Sci. U.S.* 54:960–967.

Ingram, V. M. 1958. How do genes act? *Sci. Am.* 198:(1) 69–74.

Jacob, I. and Monod, J. 1961. On the regulation of gene activity. *Cold Spring Harbor Symp. Quant. Biol.* 26:193–211.

Kornberg, A. 1959. Enzymatic synthesis of deoxyribonucleic acid. *The Harvey Lectures* 53:83–112. The Harvey Soc. of New York, New York: Academic Press, 1959.

Lederberg, J. 1960. A view of genetics. *Science* 131:269–275.

Light, S. F. and Weesner, F. M. 1951. Further studies on the production of supplementary reproductives in Zootermopsis (Isoptera). *J. Exptl. Zool.* 117:397–414.

Luscher, M. 1953. The termite and the cell. *Sci. American.* 188:74–78.

Lwoff, A. 1953. Lysogeny. *Bact. Rev.* 17:269–337.

McAuslan, B. R. 1963. The induction and repression of thymidine kinase in the Poxvirus-infected HeLa cell. *Virology* 21:383–389.

McClung, C. E. 1901. Notes on the accessory chromosome. *Anat. Anz.* 20:220–226.

———. 1902. The accessory chromosome—sex determinant? *Biol. Bull.* Woods Hole 3:43–84.

Meselson, M. and Stahl, F. W. 1958. The replication of DNA in *Escherichia coli. Proc. Natl. Acad. Sci. U.S.* 44:671–682.

Mirsky, A. E. and Pollister, A. W. 1943. Studies on the chemistry of chromatin. *Trans. N.Y. Acad. Sci. Series II* 5:190–198.

Monod, J. and Jacob, F. 1961. General conclusions: Teleonomic mechanisms in cellular metabolism, growth, and differentiation. In Cellular regulatory mechanisms, *Cold Spring Harbor Symp. Quant. Biol.* 26:389–401.

Moore, J. A. 1963. *Heredity and Development.* 245 pp. New York: Oxford University Press.

Morgan, T. H. 1910. An attempt to analyze the constitution of the chromosomes on the basis of sex-limited inheritance in *Drosophila. J. Exptl. Zool.* 11:365–412.

Muller, H. J. 1927. Artificial transmutation of the gene. *Science* 66:84–87.

———. 1950. Some present problems in the genetic effects of radiation. *J. Cell. Comp. Physiol.* 35:9–70.

Nirenberg, M. W. 1963. The genetic code II. *Sci. American* 208:(3) 80–94.

Nirenberg, M. W. and Matthaei, J. H. 1961. The dependence of cell-free protein synthesis in *E. coli* upon naturally occurring or synthetic polyribonucleotides. *Proc. Natl. Acad. Sci.* U.S. 47:1588–1602.

Ochoa, S. 1963. Synthetic polynucleotides and the genetic code. In Informational macromolecules. *Rutgers Inf. Symp.* eds. H. J. Vogel, V. Bryson and J. O. Lampen. pp. 437–449. New York: Academic Press.

Painter, T. S. 1934. Salivary chromosomes and the attack on the gene. *J. Hered.* 25:464–476.

Pardee, A. B., Jacob, F. and Monod, J. 1959. Specific inhibition in beta galactoside genes. *J. Mol. Biol.* 1:165–178.

Pauling, L., Itano, H. A., Singer, S. J., and Wells, I. C. 1949. Sickle cell anemia, a molecular disease. *Science* 110:543–548.

Pollister, A. W., Swift, H. and Alfert, M. 1951. Studies on the desoxypentose acid content of animal nuclei. *J. Cell. Comp. Physiol.* 38 (Suppl. 1): 101–119.

Ptashne, M. and Gilbert, W., 1970. Genetic repressors. *Sci. Am.* 222(6): 36–44.

Reiner, J. 1968. *The Organism as an Adaptive Control System,* 224 pp. Englewood Cliffs, N.J.: Prentice-Hall.

Rose, S. M. 1952. A hierarchy of self-limiting reactions as the basis of cellular differentiation and growth control. *Amer. Nat.* 86:337–354.

Spirin, A. S. 1966. On "masked" forms of messenger RNA in early embryogenesis and in other differentiating systems. *Current Topics in Dev. Biol.* 1:1–38.

Stedman, E. and Stedman, E. 1950. Cell specificity of histones. *Nature* 166:780.

Sturtevant, A. H. 1913. The linear arrangement of six sex-linked factors in *Drosophila,* as shown by their mode of association. *J. Exptl. Zool.* 14:43–59.

Sutton, W. S. 1902. On the morphology of the chromosome group in *Brachystola magna. Biol. Bull.* Woods Hole 4:24–39.

———. 1903. The chromosome in heredity. *Biol. Bull.* Woods Hole 4:231–51.

Swift, H. 1950. The desoxyribose nucleic acid content of animal nuclei. *Physiol. Zool.* 23:169–198.

Sypherd, P. S. and Strauss, N. 1963. The role of RNA in repression of enzyme synthesis. *Proc. Natl. Acad. Sci.* U.S. 50:1059–1066.

Taylor, J. H., Woods, P. S. and Hughes, W. L. 1957. The organization and duplication of chromosomes as revealed by autoradiographic studies using tritiumlabeled thymidine. *Proc. Natl. Acad. Sci.* U.S. 43:122–128.

Tyler, A. 1967. Masked messenger RNA and cytoplasmic DNA in relation to protein synthesis and processes of fertilization and determination in embryonic development. In *Control Mechanisms in Developmental Processes.* ed. M. Locke, pp. 170–226. New York: Academic Press.

Umbarger, H. E. 1963. The integration of metabolic pathways. *Ann. Rev. Plant Physiol.* 14:19–42.

Vogel, H. J. 1957. Repression and induction as control mechanisms of enzyme biogenesis: The "adaptive" formation of acetylornithinase. In *The Chemical Basis of Heredity* ed. W. D. McElroy and B. Glass, pp. 276–296. Baltimore: The Johns Hopkins Press.

Watson, J. D. and Crick, F. H. C. 1953. The structure of DNA. *Cold Spring Harbor Symp. Quant. Biol.* 18:123–131.

———. 1953a. Molecular structure of nucleic acids. A structure of deoxyribose nucleic acid. *Nature* 171:737–738.

———. 1953b. Genetic implications of the structure of deoxyribonucleic acid. *Nature* 171:964.

Wilkins, M. H. F. 1956. Physical studies of the molecular structure of deoxyribose nucleic acid and nucleoprotein. *Cold Spring Harbor Symp. Quant. Biol.* 21:75–90.

Wyatt, G. R. 1952. The nucleic acids of some insect viruses. *Jour. Gen. Physiol.* 36:201–205.

Yanofsky, C. 1963. Mutational alternation of the primary structure of the A protein of tryptophan synthetase. In Informational Macromolecules, eds. H. J. Vogel, V. Bryson, and J. O. Lampen, *Rutgers Inf. Symp.* pp. 195–204. New York: Academic Press.

Zinder, N. 1958. "Transduction" in bacteria. *Sci. American* 199 (11):38–43.

———. 1963. The information content of an RNA-containing bacteriophage. In Informational Macromolecules, eds. H. J. Vogel, V. Bryson, and J. O. Lampen, *Rutgers Inf. Symp.* pp. 229–238. New York: Academic Press.

CHAPTER TWO

Spatial Tissue Relationships in the Control of Lens Regeneration in the Eyes of Salamanders

A good example of the fact that the control of regeneration is polarized has come from studies of lens regeneration in salamanders. This story begins with a publication in 1891 by Vincenzo L. Colucci. He had learned that after removal of the lens from an eye a new one could form. Four years later Gustav Wolff independently described the same process which is now known as Wolffian regeneration. What they had observed was that after lentectomy a small bit of pigmented iris tissue transformed to a perfect transparent lens. A functional lens normal in every way can arise from a very different kind of tissue in this vertebrate animal. The cells of a small part of the iris stop making pigment and start making lens proteins. As this process of detooling and retooling occurs, the cell surfaces also change and the cells no longer associate with each other to form a sheet but contact each other in such a way that the configuration has lens shape. Steps in the histological changes are given as worked out by R. W. Reyer (1954) a modern student of the problem.

The most sensitive techniques for following the retooling involve the use of antibodies. Lenses are ground and injected into rabbits. In about two weeks there is present in the rabbits' blood sera a good supply of new protein configurations—antibodies—which recognize and react with the lens substances—antigens—that incited the antibody production. Although these immune substances are best known in their role of protection from invading micro-organisms, they can also be used to trace the development of highly specific antigenic proteins.

One example of the use of antibodies to study regeneration at the molecular level is that of Tadasu Ogawa (1964). He prepared antibodies against lenses of the Japanese newt, a water salamander. He then bathed his experimental newts with slits in their corneas in diluted antilens serum for three hours. A newt with a normal eye except for the slit cornea was injured in one place only. Lenses collapsed and the cells no longer fitted together properly. The lens was the only structure affected. Immediately after lentectomy, nothing in the eye or in the rest of a test newt was affected by the three-hour bath in diluted antiserum. By 7 days after lentectomy, the dorsal part of the iris, directly above the space which the lens had vacated, was affected by the anti-lens serum. Cells of the mid-dorsal pupillary edge of the iris lost contact with their neighbors and floated off into the eye chamber. The treatment killed the cells. This indicated, as is known from studies at the molecular level, that the stretches of DNA involved in lens formation, also called lens protein genes, had become active in producing their special RNAs. These in turn had served as the codes for the antigenic lens proteins, the final products recognized by the highly specific antibodies.

The action of the antibodies against the lens-antigen-bearing-cells of the iris occurred before their black pigment had been lost. Presumably they had stopped making pigment early, but it had not yet been extruded. In general, when cells transform during regeneration they first stop doing their former jobs before taking on new ones. According to Tuneo Yamada (1966), there is a short time when the transforming tissue shows neither the old nor the new pattern of molecular activity. This loss of special function during transformation is known as *dedifferentiation*.

In Ogawa's study of the effect of antilens antibodies on regeneration of lens, there came a time when the antibodies destroyed both the small lens still attached to the iris and the surrounding iris tissue. Such an animal could still make a new lens after the antibody treatment. The cells which took on the job were the cells of the iris just beyond those which had been killed. A regenerating system always works this way. There are always reserves ready to transform if those ahead fail.

Finally, after the new lens had formed and had been detached from the iris, the antibodies damaged only the lens. Those iris cells which could still have become lens if needed settled back to being iris. No part of the iris could then be affected by lens antibodies.

Lens antisera have also been used to learn where lens antigens form during embryonic development. As expected, they interfere with the differentiation of lens, but they also destroy the nearby optic vesicles, especially that part becoming neural retina. Adjacent brain regions and the ectoderm adjacent to the lens which is about to become cornea are also targets for lens antibodies.

A modification of the antigen detection method yields more graphic results. Antibodies against lens can be coupled with fluorescent dyes and used on microscope slides of developing embryos. The fluorescent dye carried to the antigen by the specific antibody makes visible the sites of the antigen. Another way to track antigens is to apply an unstained antibody against a purified antigen to the microscope slide and then make the complex visible with a fluorescent antibody against the first antibody. By such methods it was possible to demonstrate that during development of the eye special lens proteins, crystallins or similar proteins, can be detected not just in lens but in the adjacent tissues, retina, brain and cornea (ten Cate and van Doormaalen, 1950; Fowler and Clarke, 1960a and b; Langman, Maisel and Squires, 1962). The same or similar antigens may even be present in the unfertilized egg (Flickinger and Stone, 1960). However, there comes a time in development when antibodies against a highly purified lens protein, alpha crystallin, will react only with lens (Ikeda and Zwaan, 1967).

One is led to think by this work and other work with antibodies that as embryonic differentiation proceeds, cells become more and more specialized but that earlier, as in the case of the lens proteins, some cells in adjacent regions are duplicating lens activity. In fact, only when differentiation is completed does only lens do lens work. Even then, after removal of lens, the neighbors in a certain position with respect to the lens can transform to lens.

The idea has been advanced several times in several forms that differentiation may be the result of an antigen-antibody reaction between neighboring cells. In its simplest form, the antigen configuration on the surface of one cell was thought to call forth a different but complementary antibody-like configuration in its neighbors. This has been an appealing hypothesis, but there is no clear evidence in its favor and it is contradicted by evidence presented later.

Very soon after the demonstration of Wolffian regeneration in salamanders investigators began to ask why the iris transformed. What started it? Various kinds of stimulation were suspected. It seemed possible that injury to the lens or its neighboring tissues might be the stimulus. Injury accompanying lentectomy was not the cause, because if a lens were removed with the attendant injury and then reinserted in the proper position in the eye, a new lens did not form (Fischel, 1902; Mikami, 1941). It was thought that possibly anything occupying the lens position would act like lens. This was found not to be so by Fischel (1902) and by Wachs (1914). Beads of various substances, for example, potato, were inserted in place of lens, but a lens usually regenerated. Bits of living salamander tissues were left floating in the eyechamber after lentectomy, but only lens acted to prevent the transformation of iris to lens. The control was quite specific.

It appears that only a large lens is capable of preventing or delaying the Wolffian transformation while floating free in the eye chamber. This had been considered an indication that older, larger lenses were qualitatively different from younger, non-suppressing lenses. However, David Frost (1961) has shown that several smaller lenses, equal in weight to a larger one, are just as effective in suppressing lens regeneration. It appears that the smaller and larger lenses are equally effective per unit of mass. Only a large one can by itself work across a fluid gap.

In general tissues can control the differentiation of others only when there is good cellular contact. Even in the eye during normal development this seems to be the case (Goro Eguchi, 1961). Eguchi worked with larval newts. When a small lens was detached, although it remained in the eye, some dorsal iris cells transformed to lens. The young lens of these small salamanders had to make cellular contract before it could prevent Wolffian transformation.

Although the adult salamander lens can work across a fluid gap, it is much more effective when in contact with the iris which it is preventing from transforming to lens. Yoshiki Mikami (1941) reviewed the evidence and pointed out several ways in which this had been learned. For example, Wachs (1914) had replaced lenses with slightly smaller ones. When the substitute was perfectly positioned and touched the iris, it completely suppressed the Wolffian transformation. When only slightly displaced so that there was a small gap between substitute lens and iris, the iris could form a lens vesicle but was stopped before lens fibers appeared. Aldo Spirito and Giacinto Ciaccio (1931) pushed tiny glass cylinders between lens and the dorsal iris and found that the dominance of lens over iris was blocked. A tiny lens could arise in the proper dorsal position. Mikami himself showed in a number of cases that young lenses could not block the transformation unless cellular contact was made between lens and iris.

In all organisms there is a right spot for a given structure. In many organisms if that structure is removed, adjacent cells will transform to the missing structure. These simple direct experiments and others like them mostly by H. Wachs in the second decade of this century, followed by those of Tadao Sato and Leon Stone and their associates, established an important relationship. Many cells, possibly all, lying near the lens position are potentially lens. They can express this potentiality only in the absence of lens or in some cases when it is present but detached.

The first explanation of the prevention by lens of other nearby cells from achieving lens status was given by the great German embryologist, Hans Spemann. He and Warren Lewis, both in 1904, had just demonstrated the phenomenon of induction in the amphibian eye (rev. by Spemann, 1938). The eye begins as a lateral outgrowth from the embryonic brain. When it reaches the surface of the head, a lens forms from the outer layer or ectoderm. Both had shown a causal connection in this relationship of eyecup to surface ectoderm. Lewis, by moving an eyecup into the trunk region under cells about to become epidermis, saw that portion of the presumptive epidermis over the eyecup become lens. Spemann had learned of lens induction by moving presumptive trunk skin to the side of the head over the eyecup. It was clear that eyecup was changing the adjacent surface of the embryo so that it became lens.

Spemann learned something else. Early in development, any small group of surface cells can be transplanted to the region over an eyecup, and they will become a whole lens or part of a lens, whatever is proper for the position. As development proceeds, the region of potential lens tissue shrinks. Finally only those cells which are actually forming lens are capable of it. Spemann realized that both in embryonic development, and later during regeneration, as some cells differentiate into lens, they prevent their neighbors from doing the same thing.

Spemann (1905) wrote of inhibitory control only one year after his first induction studies, and he tied the two together. It was assumed that the eyecup was transmitting something which would today be called an inductor to the surface cells. These, as they were transformed by the inductor, soaked it up and kept it away from other potential lens cells.

Doubt that inhibition results from competition for a scarce inductor arose from studies in which it was shown that under certain conditions more than one lens can form. This means that if there is a lens inductor, there.is more than enough for one lens. The conditions under which more than one can arise are instructive. Wachs in 1914 and many who followed him have reported that if a piece of iris is set afloat in the aqueous humor of a lensless eye, it will form a lens. In addition, the intact mid-dorsal rim of the iris of the host eye will form lens. In fact, many lenses can form from one iris, provided it is cut into many pieces. Isolation from other potential lens-producing regions is the necessary condition for the production of more than one lens.

The important question now is, "Why after lentectomy does only one spot in the iris transform to lens?" Sato (1930) and Mikami (1941) demonstrated that when small pieces of iris were transplanted to lensless eyes parts of many of them transformed to lens. There was a very clear regional difference in ability to effect the iris to lens transformation. In Mikami's experiment there was successful transformation in 84 percent of the mid-dorsal pieces. The ability to do this graded off in all directions. Those far out from the center near the lateral equatorial region transformed in only 15 percent of the cases. Test pieces from the ventral region did not transform at all in these experiments. These isolation tests of potency to form lens demonstrated that even in an adult animal the entire upper half of the iris is potentially lens. It is noteworthy that this mid-dorsal region of the iris, which forms lenses faster and more nearly perfectly when isolated, is the only region to form lens when the iris is intact. This is the phenomenon of dominance. It is encountered in all developing organisms, but a more thorough analysis of it has been made in other places than in the amphibian eye. It will be considered in greater depth in later chapters. It should also be noted here that although this small body of work was in itself not the basis for the gradient theory of development proposed by Charles Manning Child (1929), similar work on a great variety of organisms did lead to that theory.

Something has been learned of the mechanism of this control in the iris. First of all, the system over which the control moves is highly polarized. When slits were made in an iris of a lensless eye, and if those slits remained open, not only the dorsal rim produced a lens, but the tissue on the far side of the slit also produced a lens (Stone, 1954). Dominance did not jump the narrow liquid gap. Cellular contact seems necessary for this control. When contact is broken, any part of the dorsal iris can transform. Cellular contact is not needed when a full-grown lens is suppressing transformation. It can be floating free and is still effective as noted above. At the time when the mid-dorsal area is dominating the rest of the iris in the early stages of regeneration, the amount of material involved should be less. The amount of controlling substance may be so little that only when there is cellular contact is there enough reaching the cells for control to be exerted.

The thought has arisen here again just as it did in connection with lentectomy that the act of cutting a slit might release substances called *wound hormones* which might stimulate regeneration. This is a natural hypothesis, and it has been suggested hundreds of times as the stimulus to regeneration in all kinds of organisms, but there are no demonstrations of its validity. In this case as in the others there is a simple way of learning that the gap rather than the cutting is important. If several slits are made with some allowed to heal and others held open, as for example with plio-film (Stone, 1954), only those which remain open break the dominance and allow accessory lenses to regenerate.

This gap does not have to be a true physical gap. Yô K. Okada (1935) cut slits in iris after lentectomy and inserted bits of cornea in the slits. The cornea was just as effective in breaking the dominance as was a fluid gap. Most of the time the accessory lens was on the far side of the corneal graft away from the dorsal pupillary rim where the larger, more rapidly forming primary regenerate lay. One might think of the cornea as a physical barrier to the passage of some substance. However, when the corneal graft was placed far back at the junction between iris and retina, a secondary lens could arise on the iris side with no break in the iris between the position of the primary and secondary regenerates. How the control of the primary over the secondary center of regeneration failed in these cases has not been determined for the eye. We shall be coming to other organisms where there is greater understanding of how the polarity of controlling systems can be blocked or inverted.

For some time now it has been thought that both stimulating and in-hibitory substances are used as controlling agents. A number of experiments indicate that retina is needed by iris during transformation to lens. For example, as first learned by Wachs (1914), iris transplanted to a foreign region of the body does not transform to lens unless it is accompanied by retina. Even when the eye is in its normal position, an artificial barrier inserted between retina and iris will sometimes prevent lens regeneration (rev. by Stone, 1959). The interpretation has been that the retina was releasing an inductor of lens to the iris. Since the retina is one of the descendants of the eyecup this seemed natural. However, as more has been learned it has become quite clear that the induction of lens is not as simple as some of the early hypotheses suggested. Induction was thought to result from an inductor from a single tissue—in this case eyecup or retina.

We know now that both during early development in the embryo and during regeneration later a perfect lens can arise only when the proper surrounding structures are in normal position. During regeneration, a piece of iris can become a normal lens in only one position, the normal position. If the transplant comes to lie behind the lens position, it will transform to a lens with too much fibrous material and too little epithelium.

Forward of the normal lens position, a lens with the reverse abnormality will arise, relatively too much epithelium (Mikami, 1941).

The statement that any one tissue or structure induces another tissue or structure is probably an oversimplification. There are changes occurring in the surface ectoderm all over an early embryo. In very early stages, as Spemann (rev. 1938) demonstrated, presumptive ectoderm from any part of the embryo can become lens if put in the proper place on the side of the future head. As time goes on, the area of lens potentiality grows smaller and smaller. Finally, as the lens is forming, no other area is capable of becoming lens. Changes are occurring all over the body, and as ectoderm differentiates elsewhere, it loses its ability to become lens. Apparently during differentiation ectoderm in the rest of the head is more like that of lens than is trunk ectoderm. As Spemann demonstrated, there is a time during shrinkage of the competent area when trunk ectoderm can no longer become lens but non-lens head ectoderm can still become lens. This kind of process is quite general, as Charles Waddington (1956) has pointed out. It is quite clear now that the pathways of differentiation become more in number as differentiation proceeds. The cells lying close to each other progress along a common pathway for a longer time than do two groups of cells lying farther from each other.

We have swung away from the notion of one tissue or structure acting as an organizer and directing the fate of the cells lying around it. In the eye, this has been shown by several investigators. One particularly enlightening experiment was that of Okada and Mikami (1937). It was already known from the work of Spemann (rev. 1938) that in some amphibian species the eyecup must touch surface ectoderm for lens to form. In other species, lens forms even if one prevents the eyecup from reaching the surface. In these species the control spreading through the ectoderm and other non-eyecup tissues was sufficient. Okada and Mikami working with the Japanese newt were using a species in which lens formation did not occur if the eyecup was removed. However, a lens could form after various other embryonic tissues were inserted in the void left by the extirpated eyecup. It appears that the cells in the position where lens was to form had already been affected by their neighbors and had already proceeded far along the pathway to lens when the eyecup was extirpated. In some species, extirpation of the eyecup would not have prevented the further differentiation into recognizable lens. It is most interesting that Okada and Mikami could get their embryos to produce lens if they packed the void left by the extirpated eyecup with any of a variety of embryonic tissues, ectodermal, mesodermal, or endodermal. The best substitute for eyecup was the adjacent brain. But again only eyecup in position beneath lens led to perfect lens formation. The foreign tissues may not have given any specific lens information but may have acted to pass information from other nearby tissues. Obviously, lens information was already there before eyecup was removed. The dependence of a given tissue during its differentiation on more than one neighbor is shown clearly in the work of Antone Jacobson (1958). In this work, bits of presumptive lens were cultured alone or in combination with one or more neighboring tissues. Alone, pieces could not make the last steps to lens. As the number of neighbors increased, the incidence of lens development increased.

All of these things taken together indicate that the pathway of differentiation of a cell is influenced by many neighbors. We shall see later that events far away may determine what differentiates in a given spot.

So far, we have considered only transformation of dorsal iris to lens in the adult and the formation of an embryonic lens from surface ectoderm. These are the usual sources, but other eye tissues can also transform to lens. Reyer (1950) has demonstrated that if a foreign, non-reacting ectoderm is transplanted over an embryonic eyecup, part of the developing iris will transform to lens in the embryo just as it does during regeneration in the adult.

It is usual, once an iris has formed, for it to be the tissue to regenerate lens after lentectomy. In the clawed toad, *Xenopus*, the tissue which transforms to lens is the cornea (Freeman, 1963; Campbell, 1965).

In many Amphibia, isolated pieces of the pigmented portion of the retina can form small incompletely developed lenses known as lentoids.

One gains the impression from all of this work that many of the cells in an amphibian eye are capable of producing lens, but that some lie in a favored position. In an embryo, this position is in the surface ectoderm just over the eyecup. Once it has started to become lens, other potential lens areas are suppressed. If, as in the experiment by Reyer (1950), foreign ectoderm is in the proper place, a lens does not form there, and lens formation is not suppressed in the mid-dorsal iris. If this area is rendered incapable of forming lens by replacing it with foreign tissue, as in an experiment by Stone and Vultee (1949), iris both medial and lateral to the foreign graft are not suppressed and both form lens. In this case, with foreign tissue between them neither can suppress lens differentiation in the other. If no iris is present or if connections have been cut, parts of the retina can become lentoids. These, completing their development with the wrong neighbors, are quite abnormal.

In some amphibians Sato (1935) has demonstrated that iris can become not only lens but also retina. If pieces of iris of these species are placed in lensless eyes, there is some transformation to lens and some to retina. If, however, the same operation is performed but the lens is left in place, or if one has regenerated sufficiently, the iris implants continue to produce retina with the same frequency but do not produce lens.

Iris normally lies between retina and lens. In these species it can transform in either direction. This is the first instance of what appears to be generally true. A structure can be inhibited by materials which normally come from one direction but still develop in the other direction.

The evidence above indicates the possibility that each tissue as it becomes lens may produce inhibitors which move to adjacent cells, preventing them from transforming to lens. Stone and Vultee published an abstract in 1949 in which they told of their test for an inhibitor of lens regeneration. The Wolffian transformation was compared in two sets of lensless eyes. One set received a daily injection of humor from eyes with lenses. The other set was injected with an innocuous salt solution. The work was continued and published in full later. The cases were few and there were a number of failures (Stone, 1963), but sometimes lensless eyes receiving injections from eyes with lenses did fail to regenerate as long as the injections were continued and started after the injections were discontinued. The results were tantalizing but not conclusive.

When one removes a lens from an eye, the first changes can be observed the next day in the iris as part begins to transform to lens. If the lens does suppress by substances it produces, these must decrease in concentration within a day below the threshold for suppressing. One could not expect by taking a fraction of the humor from an eye with lens to have enough inhibitory material at that one instant of time to last for 24 hours.

Stephen Smith (1965) approached the problem with this in mind. He homogenized lenses of adult newts and subjected the materials to starch electrophoresis at a pH of approximately 8.6. The starch slabs were cut in half in such a way that each half had the same materials. One was stained for protein, and the bands appeared. The slabs were then laid side by side and small pieces of starch taken from the unstained slab at the level of all the bands and beyond the bands. These pieces were then implanted individually in lensless eyes. This was a way of testing the effect of the materials in each band. One could expect that there would be a slow seepage of the materials from the starch. All of it would not be added to the aqueous humor immediately where it might have been carried away or enough inactivated so that the quantity was diminished below an effective level.

The material of two band levels contained lens-inhibiting substances. The slowest-moving negatively charged band and the most rapidly moving positively charged protein bands contained lens-inhibiting material. The active material may be protein or something which moved at the same rate. The fact that trypsin removed inhibition indicated that the protein itself was inhibitory.

This is the extent of our understanding of control by repression in the amphibian eye as a result of this approach. Even in the adult newt, many eye tissues can climb the ladder to lens, but the presence of others on higher rungs prevents further upward progress. Repression seems to pass down the ladder.

This kind of study and similar studies have been made on a variety of organisms. We shall consider some of them in the remaining chapters and search for general principles.

References Cited

Campbell, J. C. 1965. An immuno-fluorescent study of lens regeneration in larval *Xenopus laevis. J. Embryol. Exptl. Morph.* 13:171–179.

Child, C. M. 1929. Physiological dominance and physiological isolation in development and reconstitution. *Arch. Entw-mech. Org.* 117:21–66.

Colucci, V. L. 1891. Sulla rigenerazione parziale dell' occhio nei tritoni. Isto-genesi e svilluppo. Studio sperimentale. *Mem. Accad. Sci. Ist. Bologna Sez. Sci. Nat.* (5)1:167–203.

Eguchi, G. 1961. The inhibitory effect of the injured and displaced lens on the lens formation in *Triturus* larvae. *Embryologia* 6:13–35.

Fischel, A. 1902. Weitere Mittheilung über die Regeneration der Linse. *Arch. Entw-mech. Org.* 15:1–138.

Flickinger, R. A., and Stone, G. 1960. Localization of lens antigens in developing frog embryos. *Exptl. Cell. Res.* 21:541–547.

Fowler, I., and Clarke, W. M. 1960a. Development of anterior structures in the chick after direct application of adult lens antisera. *Anat. Rec.* 136:194–195.

———. 1960b. Inhibition of proamnion and amniotic head field with lens antisera. *Anat. Rec.* 137:357–358.

Freeman, G. 1963. Lens regeneration from the cornea in *Xenopus laevis. J. Exptl. Zool.* 154:39–65.

Frost, D. 1961. Inhibition of lens regeneration by implanted lenses in the eye of the adult newt *Diemictylus* (= *Triturus*) *viridescens*. *Devel. Biol.* 3:516–531.

Ikeda, A., and Zwaan, J. 1967. The changing cellular localization of α-crystallin in the lens of the chicken embryo, studied by immunofluorescence. *Devel. Biol.* 15:348–367.

Jacobson, A. G. 1958. The roles of neural and non-neural tissues in lens induction. *J. Exptl. Zool.* 139:525–557.

Langman, J., Maisel, H., and Squires, J. 1962. The influence of lens antibodies on the development of lens antigen containing tissues in the chick embryo. *J. Embyrol. Exptl. Morph.* 10:178–190.

Lewis, W. 1904. Experimental studies on the development of the eye in Amphibia. I. On the origin of the lens. *Rana palustris. Am. Jour. Anat.* 3:505–536.

Mikami, Y. 1941. Experimental analysis of the Wolffian lens regeneration in adult newt, *Triturus pyrhogaster. Jap. J. Zool.* 9:269–302.

Ogawa, T. 1964. The influence of lens antibody on lens regeneration in the larval newt. *Embryologia* 8:146–157.

Okada, Y. K. 1935. The lens potency posterior to the iris. *Proc. Imp. Acad.* Tokyo 11:115–118.

Okada, Y. K., and Mikami, Y. 1937. Inductive effect of tissues other than retina on the presumptive lens epithelium. *Proc. Imp. Acad.* Tokyo 13:283–285.

Reyer, R. W. 1950. An experimental study of lens regeneration in *Triturus viridescens viridescens*. II. Lens development from the dorsal iris in the absence of the embryonic lens. *J. Exptl. Zool.* 113:317–353.

———. 1954. Regeneration of the lens in the amphibian eye. *Quart. Rev. Biol.* 29:1–46.

Sato, T. 1930. Beiträge zur Analyse der Wolffschen Linsenregeneration. *I. Wilhelm Roux' Arch. Entwicklungsmech. Organ.* 122:451–493.

———. 1935. Beiträge zur Analyse der Wolffschen Linsenregeneration. *Arch. Entw-mech. Org.* 133:323–348.

Smith, S. D. 1965. The effects of electrophoretically separated lens proteins on lens regeneration in *Diemyctylus viridescens. J. Exptl. Zool.* 159:149–166.

Spemann, H. 1905. Uber Linsenbildung nach experimentellen Entfernung der primaren Linsenbildungzellen. *Zool. Anz.* 28:419–432.

———. 1938. *Embryonic Development and Induction*. 401 pp. New Haven, Conn.: Yale University Press.

Spirito, A., and Ciaccio, G. 1931. Ricerche causali sulla rigenerazione del cristallino nei tritoni. *Boll. Zool. Napoli* 2:1–7.

Stone, L. S. 1954. Lens regeneration in secondary pupils experimentally produced in eyes of the adult newt, *Triturus v. viridescens. J. Exptl. Zool.* 127:463–492.

———. 1959. Regeneration of the retina, iris and lens. In *Regeneration in Vertebrates* (C. S. Thornton, ed.) pp. 3–14. Chicago, Ill.: Univ. Chicago Press.

————. 1963. Experiments dealing with the role played by the aqueous humor and retina in lens regeneration of adult newts. *J. Exptl. Zool.* 153:195–210.

Stone, L. S., and Vultee, J. H. 1949. Inhibition and release of lens regeneration in the dorsal iris of *Triturus v. viridescens. Anat. Rec.* 103:144–145.

ten Cate, G. and van Doormaalen, W. J. 1950. Analysis of the development of the eye-lens in chicken and frog embryos by means of the precipitin reaction. *Proc. K. Ned. Akad. Wet.* 53:894–909.

Wachs, H. 1914. Neue Versuche zur Wolffschen Linsenregeneration. *Arch. Entw-mech. Org.* 39:384–451.

Waddington, C. H. 1956. *Principles of Embryology*, 510 pp. New York: Macmillan.

Wolff, G. 1895. Entwicklungsphysiologische Studien. I. Die Regeneration der Urodelenlinse. *Wilhelm Roux' Arch. Entwicklungsmech. Organ.* 1:380–390.

Yamada, Tuneo. 1966. Control of tissue specificity: The pattern of cellular synthetic activities in tissue transformation. *Am. Zool.* 6:21–31.

CHAPTER THREE

Polarized Control of Regeneration in the Amphibian Limb

We are going to be looking at the same problems in the limb that have been treated in the eye. The story begins with the demonstration by the Abbé Spallanzani (1769) that newts can regenerate functional limbs. It has since been learned that many of the tailed Amphibia can regenerate just about any part if the removal of that part does not kill before regeneration can take place. Tails, eyes, major parts of the digestive tract and other internal organs, parts of the head—all can be regenerated.

The first problem to be dealt with arises from the fact that ordinarily only the missing part is regenerated. What could the nature of the system be that can respond to a loss by replacing the lost part? When one segment of a finger of a newt is removed, just the one segment is replaced. If a limb is amputated through the wrist, it is only the part beyond or distal to the wrist which is made anew. No matter what the level of amputation along a limb, only the more distal parts are replaced. The stump plus regenerate add up to a whole limb. Under ordinary conditions there is never too little or too much. The process in larval amphibians requires about 15 days. It may take twice or three times as long in an adult. After that, the growth of the new parts requires several more weeks or months before the size of the regenerate closely approaches that of the normal limb.

One of the first clues concerning the nature of regeneration in the limb came when Oskar Kurz (1922) reversed the polarity of the distal part of a salamander's limb stump. This was done by cutting out the knee region, turning it 180°, and grafting it by its original distal end to the limb stump. Sometimes the reversed grafts were made to the side of the body. In all of these cases the free amputation surface from which regeneration could be expected was on the lower thigh. Facing as it did, it might have been expected to complete itself by producing an upper thigh and maybe even the rest of a salamander. This never happened. Instead, the structures were always lower parts of limbs, with feet ending in toes.

Ludwig Gräper (1922) did the same kind of reverse transplantations, as he called them, with the hind limbs of tadpoles. Figure 3-1 shows the result of one of his transplantations. After transplantation, there had been upper thigh still attached with normal polarity to the body; then, reversed 180°, came upper part of lower leg, knee and lower part of thigh. The first part of the regenerate was a lower thigh just like the one it regenerated from but it faced in the opposite direction. Then followed, in normal orientation with respect to the body, another knee, a lower leg, and a foot.

At first this is confusing. The limbs were regenerating some duplicate parts. From the experiments on lens regeneration one learns that a lens already present prevents nearby tissues from becoming lens. What is different about the limb?

Hans Przibram, a Viennese professor, and his students experimented and began asking the questions which have led to our present level of understanding. The relationship was demonstrated in another way by several workers: Flat-Hauser and Przibram (1930); Przibram (1931), and by Chen (1933). They left limbs of salamanders intact, except for the removal of either the distal half or the proximal half of the long bone. The proximal halves regenerated distal halves, and the bones were completed. When only distal halves were left in place, the regenerate which formed and grew proximally was not a proximal half but a distal half. The original distal half left in place and the regenerate from it were mirror images of each other. Przibram posed the question in a lecture in London (1931). He said, " In short, distal parts do not assume the character of the more basal. This assertion may sound trivial to you: if regeneration means restoration of a lost part, why should we expect anything else differentiating from a given level than the missing?"

What happens when regeneration occurs from a reversed stump has been shown in another way. Several contemporary students of regeneration, especially Elmer Butler (1955), James Dent (1954), and K. W. Oberheim, and Wolfgang Luther (1958) have grafted limbs so that regeneration can occur in both the distal and proximal direction from the

EXPERIMENT

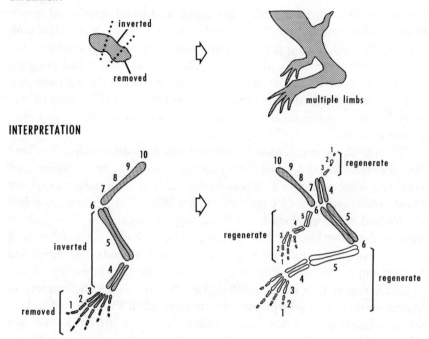

INTERPRETATION

FIG. 3-1.   REGENERATION OF MORE DISTAL STRUCTURES AFTER INVERSION OF PART OF A DEVELOPING LIMB [*adapted from Ludwig Gräper (1922)*].

The distal third of the limb was removed, and the middle third was inverted. Regeneration occurred from three surfaces: from the new distal end of the inverted piece and from two points where the old distal end of the inverted piece joined the upper leg. After the skeleton was studied, it became clear that each surface produced more distal structures. If limb levels are numbered from 1 at the distal ends of the toes to 10 at the shoulder, the inverted section might be 4, 5, 6. After inversion and amputation, the sequence was 10, 9, 8, 7, 4, 5, 6. From 6 at the distal surface of amputation 5, 4, 3, 2, 1 regenerated. The sequence in the main axis after regeneration was 10, 9, 8, 7, 4, 5, 6, 5, 4, 3, 2, 1. It is clear that regeneration is not a case of simply replacing lost parts. In this case 5 and 4 were regenerated when the same levels were part of the stump. Instead the distal regenerate, as well as the two lateral regenerates, contain all of the levels normally lying distal to the regenerating level.

same leg. One method, that of Butler, involved sewing a cut distal end of an arm into the side. The result was a limb attached normally at the shoulder and artificially at its distal end. After nerves and blood vessels had grown in and made connections at the attached distal end, the doubly serviced limb was cut through the upper arm. One stump, the one attached at the shoulder, was oriented with proper polarity. The other had reversed polarity. Both were cut at the same level by the single isolating cut. Although the direction of orientation of structures in these stumps was opposite, the regenerates on them arose as identical structures and with proper polarity.

The Oberheim and Luther experiment was done differently. They bent hind legs ventrally and sewed them into place across the pelvic region after removing some of the skin. These limbs received a secondary supply of blood vessels and nerves by their attached middles. Later, when secondary service had been established, the limbs were amputated to either side of their mid-connection. In this way, regeneration could occur from both ends of a single stump. Feet regenerated from both ends. Oberheim and Luther emphasized that limb regeneration is the transformation of any proximal region to a more distal region. We are to see in a variety of organisms that any level will transform to more distal levels, provided these are not already present distal to it. This work also points up the fact that whether the stump has an originally distal or proximal surface is immaterial. This fact is important because it eliminates any theory of pattern formation based on the simple completion of the missing part of the pattern by spread of pattern from old organized to new unorganized tissue. This is true because the patterns of old and new are reversed. Also, any kind of freely diffusing specific inhibitor is ruled out. A regenerating surface produces more distal structures regardless of the presence or absence of those structures a short distance behind the amputation surface.

Under what conditions do limb levels transform to more distal levels? In a normal intact limb there is rigid control not allowing an upper arm to transform to a hand or forearm. Knowledge concerning this was contributed by J. Faber (1960). He amputated large parts of limbs of salamanders and allowed the stumps to regenerate until the new regions were clearly delimited and new skeletal structures and joints were forming. He then cut off the regenerates and divided them into forearm and upper arm parts. These were grafted separately with normal orientation onto backs. There they continued to develop but did not grow appreciably. The distal halves produced hands and forearms as expected. The proximal halves—already with small humeri at the time of transplantation—transformed and also produced hands and forearms. On the back where the amount of limb material could not increase, the inherent tendency to transform to more distal structures was strikingly evident. Because of the small initial amount of material and its failure to increase, all proximal structures were used up during some of the transformations. Here we see the tendency of proximal structures to become more distal. Why do they fail to transform in an intact limb?

There has been a start made in understanding the nature of the control in the amphibian limb. N. V. Nassonov (1930) and W. Kasanzeff (1930) discovered a very simple way to cause the proximal part of a limb to transform to distal structures in an intact limb. All they did was tie threads around the limbs of axolotls—rather large Mexican salamanders—above the elbow. These threads used as ligatures were tight enough to cause some compression of the soft tissue, but not enough to completely block the flow of blood to and from the distal parts of the limbs. Weeks later, tiny regeneration buds appeared just proximal to the ligatures. Slowly the region just proximal to a ligature would transform to a hand.

The transformation of proximal to distal always occurred on the proximal side of the ligature, never on the distal. It appears that there is controlling information saying, "No hands," on the distal side of the ligature, but that it is blocked by the ligature. Possibly the ligature pressing against the tissue beneath changes it so that it does not transmit what we call morphogenetic or form-determining information. This is evidence for polarized control. Control in the limb travels from distal to proximal. We shall see later that there is also limited transmission in the opposite direction.

Paolo Della Valle (1913) had started this work of partially isolating parts of the limb. Notches which extended more than half-way across from one side to the other were made in salamander limbs. These were prevented from healing together, and a ligature was tied around the narrow bridge between the almost severed distal part of the limb and the rest of the limb. In addition, as can be seen in Fig. 3-2, the distal portion was amputated. This left three surfaces from which regeneration occurred. Usually a surface facing proximally does not regenerate, but when it does it always forms distal structures. Again we see that regeneration in the amphibian limb is always a transformation to something more distal.

Other treatments also block the transmission of morphogenetic control. Elmer Butler and Harold Blum (1955) shone ultraviolet light on the middle part of limbs. The treatment was extreme and resulted in a bad "sunburn" with much destruction and sloughing of tissue. There were cases in which

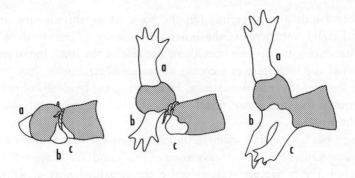

FIG. 3-2. THREE STAGES IN THE REGENERATION OF LIMBS FROM THREE NEARBY SURFACES [*adapted from Paolo della Valle (1913)*].

a: a regenerate from the distal end of the stump.

b: a regenerate from a proximal surface prevented from healing by a ligature. Although the *b* surface faced in the opposite direction it produced only distal structures.

c: a regenerate from a more proximal distal surface. It, like *a* and *b*, produced only more distal structures.

the original hand remained, and a new one grew out proximal to the destructured middle portion of the limb. X-rays can also block the flow of controlling information, but they do it more subtly without destroying tissue. As shown by V. V. Brunst, and F. H. J. Figge (1951) secondary, lateral tails can arise proximal to X-rayed portions of tails.

That X-rays can block morphogenetic control has been shown in another way. Wolfgang Luther (1948), and H. A. L. Trampusch (1959) transplanted small pieces of limb skin to tails of salamanders. Later the tails were amputated through the graft. The regenerates were tails showing no limb structure or occasionally a bit of limb structure at the side of the regenerated tail. When the same grafts were made, but to X-rayed tails, the tails did not regenerate and did not suppress limb regeneration. Small but complete limbs arose from the skin grafts (Fig. 3-3). There are two points to be noted. First, and what concerns us here, is that X-irradiation blocked the morphogenetic control by the tail. Second, and something we shall return to later, is the fact that limb skin has within it the information not just for skin but for a complete limb.

Another treatment which causes hands to grow out from sides of limbs is the transplantation of a bit of malignant tumor tissue, as done by Laurens

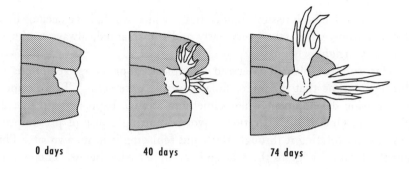

0 days   40 days   74 days

Fig. 3-3. Regeneration of limbs from limb skin transplanted to an x-rayed tail [*adapted from H. A. L. Trampusch (1958)*].

The clear patch represents normal limb skin transplanted to an X-rayed tail. The tail had been amputated at the level of the distal end of the limb skin. From the skin arose a double limb structure. The X-rayed tail did not regenerate after X-irradiation. Had it not been X-rayed, it would have regenerated the missing distal part of the tail and at the same time the limb skin would not have transformed to a limb. X-rays appear to block repressive morphogenetic control.

Ruben (1960), or by disrupting limb tissue with cancer-inducing chemicals, as done by Charles Breedis (1952). This foreign, neoplastic tissue blocks the orderly transmission of form controlling information.

ALIGNED CONTROL

Both in the eye and in the limb pieces of tissue which have been isolated tend to transform to other tissues. There is order in the direction of these transformations—toward lens or toward more distal limb structures. Lens and distal parts of a limb never transform to other tissues. They have arrived at the top of their hierarchy. These controls operating in such a way that not everything becomes lens or hand are highly aligned.

The fact of aligned control and the consequences of interfering with it in the development of the amphibian limb were demonstrated by Ross Harrison (1918, 1921). Professor Harrison systematically transplanted tissue about to become limb. Sometimes it was transplanted to limb sites, at other times to non-limb regions. Sometimes its original axes were maintained, at other times the graft was turned before being laid in place in the host site.

Presumptive limb tissue—that is, tissue which would have become limb if not moved—when taken very early and transplanted, always conforms to its new neighborhood.

Harrison transplanted older and older pieces of presumptive limb tissue during the time that the intercellular controls were acting. There came a time during development when something strange happened after transplantation. One of the operations involved cutting a disc of presumptive limb tissue, rotating it through 180°, and replacing it in its own site. This put the forward edge to the rear and the upper edge below. The antero-posterior and dorso-ventral axes of the piece were reversed with respect to the rest of the embryonic environment. These transplants of fairly young tissue developed into limbs which pointed backwards but whose upper and lower surfaces conformed with the rest of the embryo. In other words, the antero-posterior axis had been determined before the operation, but the dorso-ventral axis was fixed after the rotation. One axis had developed in the original position. The other, developing later, had developed in the new rotated position and conformed to the new environment.

The same operation performed a day later resulted in a backwards and upside-down limb. By this later time, both axes had been fixed while the tissue was still in the original position. After rotation, the new neighborhood had no effect on this already determined tissue.

John Nicholas (1924), a student of Harrison, demonstrated that the fixing of the axis in the pre-limb region was an event involving more than the future limb tissue. Nicholas left the presumptive limb tissue in place but cut out a ring of tissue around it. The ring was rotated but not the presumptive limb tissue itself. However, when limbs appeared later, their axes had been shifted by the rotation of the tissue around them. There was some indication that the control was polarized because the dorso-ventral axis was most affected by changes in position of the tissue dorsal to it.

This demonstration that control could operate along one axis at one time and along another later provided a new dimension to the study of biological organization. There is another aspect of the Harrison work which is just now being understood well enough so that it can be used in the quest of new knowledge.

In Chapter 2 the concept of polarized inhibitory control was introduced. That control is polarized and aligned, is more thoroughly known for the developing limb again largely because Harrison in his careful, thoughtful, thorough way worked through the problem.

When a presumptive limb disc was removed to another body site it could go on and produce a limb. At the same time, the gap left at the original limb site was often filled by neighboring cells during healing. These neighbors, following the removal of the primary limb-forming tissue, also became limb.

There were several observations which are important to us now. If a presumptive limb disc was removed and replaced in normal alignment, the neighbors did not form another limb. If, however, the disc was replaced at an angle to its original position, neighbors could produce an extra limb. The greater the angle of deviation—up to 180°—between presumptive limb tissue and its normal environs, the more often—up to 90 percent of the time—the neighbors produced an extra limb.

This work of Harrison and his several students tells us that controls operating during differentiation are aligned and polarized. It may be important to remember that cells can lie side by side in apparently normal union but not communicate when their alignments have been disturbed.

One feature of the controls being considered is their inhibitory nature. A limb disc properly aligned could prevent its neighbors from becoming limb. An extra limb never developed when a limb disc was transplanted to the head. The transplant could itself form a limb on the head, but it never induced an extra one in that region. This, too, may be important. These controls operate to prevent regions from doing what their neighbors are doing. They are not inducing them to do something foreign to their region. When a misaligned primary limb has a secondary limb at its base, it is simply because the inhibitory control has failed to reach the secondary limb site. In the head no degree of misalignment leads to secondary limb formation. Failure to prevent something which does not occur in the head region has no effect. There is no evidence that one limb ever tells its neighbor to form a limb. It can only tell it not to.

One observation which Professor Harrison made repeatedly did not make sense at the time. When he transplanted presumptive limb discs to a position between the fore and hind limb sites, secondary limbs often arose from the flank at the base of the transplants. It certainly appeared that the grafts were inducing non-limb region to produce limb. What did not make sense was that the extra limbs arose more often when the antero-posterior and dorso-ventral axes were unchanged. When the future limb region grafts were turned through 180°, they usually did not induce an extra limb. This was just the reverse of the behavior at the original limb site.

There are two things to be explained about this part of the work. First of all, the flank is a potential limb-producing area, and it does not need any specific limb-inducing stimulus. It is now known that many things can set off limb production in this area. Such unspecific starters as ear vesicles (Balinsky, 1925, 1926, 1927; Filatow, 1927) or celloidin beads (Balinsky, 1927) grafted under the surface or transplants of tissue dying from heavy X-ray doses (Perri, 1956) will induce this normally non-limb-producing area to produce limb. There is no need to invoke the transfer of limb-forming information from the pre-limb graft to the flank. It is difficult to conceive of a celloidin bead carrying specific limb information. The information to produce limb lies latent in the flank, and it can be brought forth by a variety of non-specific agents. We learned from the work of Harrison that all that one has to do to "induce" a limb is to change the axes of a piece of tissue with respect to its environs. Then the environs no longer receiving sufficient inhibitory messages can and do produce limb. As we shall see from work with other forms, it is very likely that all such things as celloidin beads do is change the geometry of the tissues and thereby set up new axes of control.

One interesting thing about many of these limbs induced on the flank is that their antero-posterior axes are reversed, and they are mirror images of the limbs in the normal positions. Now Harrison's paradox makes sense. He was getting more extra limbs on the flank when the presumptive limb disc was inserted with a normal a-p axis because the flank has an opposed polarity. When he turned his grafts around, he was really making their axes conform to the axes of the flank region. With conforming axes, the grafts could communicate with their neighbors and they did not produce limbs. With no conforming axes, secondary limbs and sometimes even tertiary limbs may form. The graft, as in Perri's grafts of X-rayed tissue, need not be limb grafts. Apparently any agent which interferes with the transfer of controlling information can "induce" a limb.

V. Kiortsis (1953) has found the locus at the base of the limb where the polarity changes. Another treatment which causes limbs to form out of place is the deviation of a nerve to a surface area, as first discovered by Piera Locatelli (1929). A bud or mound of tissue arises where nerve fibers make contact with epidermal cells, and if this bud lies near the base of a limb it will become a secondary limb. Kiortsis found that nerves deviated to surface points behind the forelimb region cause outgrowths which become limbs facing backwards. Up close to the limb is a line where deviated nerves induce outgrowths which become doubled limbs. One part has normal limb orientation and the other faces backward. Here again are identical or nearly identical structures arising where polarity changes.

This brings us to the question of what the nerve has to do with control. Before dealing with this question, we have a few more bits of knowledge concerning aligned control which should be considered.

Ordinarily, if a limb is amputated only one regenerate arises from the stump. If, in addition, the distal end of the stump is slit longitudinally, whole hands may form on a half stump (Weiss, 1926). Obviously there is not point for point control passing from stump to bud.

Another thing Weiss did was to cut a portion out of the side of a salamander's limbs. This left each limb with a rectangular gap along one side. Cut surfaces which faced back toward the body never regenerated anything. With greater trauma, as in Della Valle's work (1913), regenerates can arise on proximal surfaces but they produce distal structures. Those facing laterally healed and produced scar tissue, but the missing structures were not replaced from that surface. Only a surface which faced away from the body could regenerate. It could produce everything not lying distal to it.

One might think that once a bit of amphibian limb tissue is freed from distal controls, it might autonomously produce a symmetrical distal part of a limb and proceed to build more and more proximal structures until it reached a level already present. This may be true, but the stump also dictates what shall form ahead of it or at least the alignment of what is forming ahead of it. Again the work to be described was initiated by Weiss (1924). He removed hands from right and left limbs and sewed the stumps together side by side. The fused amputation surface was provided by two limbs. In many cases, doubled or partially doubled structures arose. In other cases what appeared to be single hands appeared. More detailed analysis by Alberto Monroy (1946) and by Richard Goss (1956a and b) established that the angle between the two stumps determines whether two hands or less will form. If the limb stumps lie parallel and side by side, a double structure which is composed of fused right and left hands may arise. As the angle between the two limbs increases, less and less of a regenerate appears. Finally, when the stumps are sewed together at right angles to each other, no regenerate forms. There can still be enough amputation surface exposed for two regenerates, but none appears. Monroy pointed out that if one were to draw lines parallel to the main axis of a limb, these would cross in the region where regeneration fails (Fig. 3-4).

As can be seen from the figure, wherever the hypothetical lines cross, no structures form. What could those lines be? We do not have a certain answer to this question yet, but something has been learned about them by the use of X-rays. It was noted above that X-rays block the spread of morphogenetic information. Limb regeneration fails in salamanders if the limb has received 2000–7000 Röntgen units, with the amount necessary varying with the species. Such a limb may live for years. producing new epidermis and remaining generally healthy. The X-rayed stump, however,

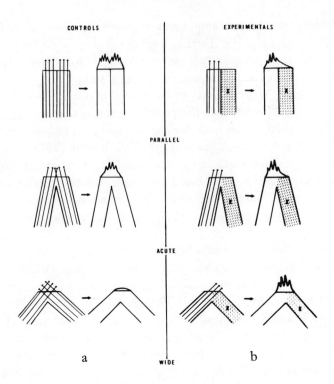

CONTROLS   EXPERIMENTALS

PARALLEL

ACUTE

a   WIDE   b

FIG. 3-4.   FAILURE OF LINEAR CONTROL IN X-RAYED LIMBS [*from Jean Ober-priller, Jour Exptl. Zool. 168, 403, 1968*)].

a: Diagrammatic representation of what happens to regenerates when stumps lie parallel, at an acute angle, or at a right or wider angle. When the hypothetical lines of control of Monroy cross no regeneration occurs. Only central regions fail to form when the angle is acute.

b: The Monroy experiment repeated but with the right limb X-rayed. In all cases the X-rayed limb failed ro regenerate and in addition failed to interfere with the regeneration of its partner.

will fail to regenerate even years later, as learned by V. V. Brunst and E. A. Scheremetjewa (1937).

The ability to regenerate is returned to an X-rayed limb by a graft of normal limb tissue. Bone and muscle can restore regenerative ability, but skin grafts result in greater restoration. The pioneers in this field were E. Umanski (1937, 1938a and b) and Charles Thornton (1942). These restored limbs can regenerate all tissues, not just more of the types contained in the normal grafts. Part of the recovery involves transformation of normal grafts to other tissues. In addition, the morphogenetic message system is restored and X-rayed tissue may also take part in regeneration (Rose and Rose, 1965, 1967).

An important clue concerning the nature of the morphogenetic control system was provided by E. E. Umanski, V. K. Tkatsch and V. P. Koudokotsev (1951). They first suppressed limb regeneration by X-rays and then grafted normal limb skin to the X-rayed stump in different orientations. Regeneration followed transplantation of rings of skin with normal orientation or turned through 180°. The latter had its distal-proximal axis reversed. Still the combination of X-rayed stump and reversed skin regenerated. The X-rayed limbs did not regenerate when they received a graft of normal skin turned through 90°. In this connection it should be recalled that regeneration can occur after removal of distal parts but not of lateral parts.

The only physical difference between the distal-proximal axis and the left-right axis which Umanski and his coworkers knew was an appreciably greater electrical resistance in the across direction. This may be the fundamental difference, as we shall see later. If morphogenetic messages were transported predominantly along the distal-proximal axis there might appear to be lines of control.

It is already clear from the work on regeneration from reversed limb segments or on regeneration of portions of bones that any level of a limb can transform to more distal structures regardless of whether the stump orientation is disto-proximal or proximo-distal. It is also true that every point in the stump is not specifying the detailed type of differentiation in the new tissue differentiating distal to it. If this were the case, whole regenerates could not form on half stumps as noted above.

Another fact which is well established is that the orientation of the regenerated structures agrees with those in the stump. Early in the analysis of how a stump controls the regeneration bud or blastema, Borivoje Milojevic (1924) transplanted very young blastemas from limb sites on one side of the body to vacated limb sites on the other side. In typical fashion, young blastemas conformed to the pattern of the new neighborhood. A young left blastema became a right limb when developing in a right limb site. Older blastemas when transplanted to foreign limb sites retained their original left-right axis. Embryologists would say the axis had already been determined. Later Günter Schwidefsky (1935) learned that there is a short period during regeneration when a blastema is partly determined. These blastemas when transplanted often produced doubled limbs. It would appear that a left-right axis was being established while the blastema was on its original site, but that the blastema was still plastic and a different axis could be imposed in part of the graft by the new site.

Another possibility might be, as in Harrison's cases, that the alignment of blastema and host site were opposed to an extent that the blastema could not suppress the host site. Then the host stump could have contributed the secondary limb. This was probably not the case in Schwidefsky's experiments because the secondary limb seemed to arise more rapidly than the host limb could have produced a regenerate. Whatever the explanation, it is clear that there is a brief period during the development of a regenerate when the influence of two sets of controls can be expressed.

Where in the stump does the ability to control axes reside? A part of the answer was provided by Milojevic. If instead of transplanting a young left blastema alone to a right site he transplanted it with a 1mm ring of skin from its stump, the axis of the new limb conformed to the axis of the accompanying old skin. For example, if the blastema plus skin was put on with its upper side down, the regenerated hand was upside down.

At about the same time that Milojevic was learning that skin could control the axis, Vera Bischler (1926) showed that discs of limb stripped of skin could regenerate when transplanted to the back. They received a covering of epidermis from the back and went on to regenerate new limbs which conformed to the axes of the stump minus limb skin. Further work indicates that probably all tissues can direct the axes of regeneration distal to them (rev. by Rose, 1964).

One can think of a stump as an organization of striving young men held down by their elders. All tissues are prepared to serve as distal tissues, but they cannot if the distal positions are filled. After removal of distal structures, the most distal of the remaining tissues are free to become the most distal tissues not forming distal to them. At the same time, various tissues of the stump can impose axes on the blastema developing distal to them.

There is much redundancy in the regenerating limb, just as, in a democracy, all individuals have a vote but some, because of their positions, have greater influence. For example, although bones have been shown to have an effect in determining the type of tissue developing distal to them (Goss, 1956a), all skeleton can be removed from a limb and still the blastema can develop into a complete regenerate including skeleton (Fritsch, 1911; Weiss, 1925; Bischler, 1926). It is clear that when bone is present it can affect the nature of the differentiation distal to it. In the work by Goss an extra ulna was implanted in forearms. After it had healed in place, the limbs were amputated through the forearm with the extra bone. In the regenerate just distal to the extra ulna an extra carpal regenerated.

Although each old structure may be shown to influence new pattern, it appears that the old pattern in the internal part of a stump may be almost irrelevant. Polejaiev (1936) dissected out major nerves and blood vessels but left them attached proximally. Next, he removed everything else inside the skin and minced the removed tissues. After they were thoroughly minced, they were packed into the skin sleeve along with the still attached nerves and blood vessels. A month later the distal end of the depatterned stump was amputated. Good, normal regenerates could form on stumps with the confused pattern. It may be that in this case the skin alone, as in Milojevic's experiments, was enough to determine the axes. The most extreme cases of control by skin were observed when Luther and Trampusch first blocked the voting power of a tail stump by X-rays, then applied limb skin. The blastema arose presumably from the epidermis and dermis and any muscle which might have been attached to the skin. This mound of tissue then formed all of the tissues of the limb in their proper places. There is no evidence here that the stump specified detailed pattern nor is there when the stump is a half stump or when most of the stump is minced muscle. Once a blastema has formed and its axes have been determined by the stump, it is an autonomous agent and tends to form a whole symmetrical distal part of a limb no matter what lies behind it.

There is something peculiar about the skin in that it seems to have more voting power than the other tissues of a stump. This is true even when the other tissues are not X-rayed. Richard Glade (1957) grafted cuffs of tail skin from salamanders to their limbs. There was a great range in the amount of tailness and limbness in the regenerates. Some of the regenerates were clearly tails even though most of the stump was limb.

The regenerates with the least amount of tailness were in some cases quite good limbs at their bases. In these cases the rest of the stump was outvoting the grafted tail skin in the proximal portion of the regenerate. Most instructive is the position of the abnormalities in these cases. There were extra cartilages up at the distal end of the regenerate, and fingers did not grow out. The new tissue closest to the old tail skin in a lateral direction could be pure limb. The greatest effect of the tail skin in the cases where it did not completely control the regenerate was at the far distal tip in some cases far removed from the skin graft.

Another time when skin is known to outvote other limb tissues is when normal tissues are used to reinstitute regeneration in X-rayed limbs. Normal skin grafts in the amputation region may lead to perfect or nearly perfect regeneration. Other normal tissues such as bone and muscle lead to quite imperfect regeneration (rev. by Goss, 1957; Trampusch, 1959; Lazard, 1959; Rose, 1964).

Two things are being considered here which may be related. First it is clear that all stump tissues have been shown to be able to control axes and/or positions of structures distal to them during regeneration. In experimental situations the skin can usually outvote the rest. The other relationship which should be considered at the same time is that distal structures can outvote proximal structures to such a degree that there is usually no contest. If distal structures are present either as the original structures or as re-generates, the same structures will not form behind them. Only when a ligature is applied or intermediate tissues are disorganized or X-rayed do distal structures fail to suppress like development in more proximal regions.

The major direction of control within the original limb or within a regenerate is from distal to proximal. The skin is an exception. It can have its effect all of the way from its position on the side of a stump up to the very distal tip where internal tissues are regenerating (Glade, 1957). Part of the reason for this may be that epidermis from old skin migrates distally and covers a stump during wound healing before regeneration begins. When regeneration begins, epidermis is the most distal tissue. With votes being passed to the internal tissues primarily in the disto-proximal direc-tion, it is no wonder that skin has so much voting power. It is probably only because of its position that it is so important. There may be nothing unique about the epidermis during regeneration except its position. We had tended to think of it as containing more useful information than other tissues. This is probably wrong. Glade (1963) has shown by grafting epidermis and dermis separately that less regional information—tail in this case—originates in the epidermis than does in the dermis, its partner forming the deeper layer of the skin. The epidermis may simply be the pathway by which information from all tissues reaches the tip of the regenerate. This requires further study.

It would seem from the points considered above that control passes to the tip around the outside by way of the epidermis. It then turns and passes proximally through the internal tissues. The first observed effect of a stump on a blastema is that of orienting. Once the axes have been established, the interactions leading to complicated structures are largely interactions between cells in the blastema. Just as an egg becomes a chicken without any old chicken parts around to tell it what to do, so, too, can a blastema on a limb become a hand by using information stored within the genetic repositories of its own cells.

There are two restrictions on the use of genetic information in a blastema: an orienting restriction, and one other. Limb tissues during regeneration are limited to becoming more distal limb tissues. They do not form noses or eyes or tails. We think of higher organisms as being composed of a number of territories. Any part of a regenerating territory can become a more distal part of that same territory.

This restriction during transformation to the use of some sub-block of genetic information of the same territory does not mean that the cells of one territory have lost the DNA used in other territories. However, once a cell has been canalized, to use the expression of C. H. Waddington, it can move forward along that canal—but not back—and then out along another canal.

It is known that cells after becoming members of one territory can, under some experimental conditions, be made to transform and serve as members of another territory. Farinella-Ferruza (1957) has transplanted embryonic tail buds to limb sites of other embryos. The tail buds of the hosts were from different genera, and they had nuclei of quite different sizes. Tail bud transplants developed into tails in limb sites. Slowly over a period of months limbs grew up at the bases of some of the tails. Some of the cells of the limbs had nuclei of host size and had arisen from a regenerating limb territory at the base of the tail. However, in time some of the former tail cells lying in position to become part of the growing limb did become limb cells. Their nuclei of graft origin had formerly been tail nuclei.

We do not know what made tail change to limb. Formerly as part of a tail it could have transformed, just as in a limb, only to more distal parts. Here it had somehow come under the influence of the limb, had lost its tailness and taken its place as part of a limb pattern. There are important questions here which remain unanswered. Did the transforming tail cells receive some specific limb commands from the limb? Was it rather a case of no longer receiving the inhibitory controls limiting it to tail level? These questions have not yet been answered in experiments on vertebrates but we shall try to answer them in later chapters on lower forms.

This brings us to the nature of the change observable at the cellular level when a region becomes something more distal. After epidermal cells migrate across the wound surface and cover it, they and the internal tissues begin to change. Cartilaginous and bony matrix dissolve and fibers disappear close under the wound surface. Muscle fibers break up into small units. Internally all cells lose all or most of their endoplasmic reticulum. Special structures such as myofibrils disappear. All cells come to look alike under both light and electron microscopes (Schotté, 1940; Thornton,

1942; Hay, 1958, 1959). This first step is dedifferentiation. The cells first stop making their special products and the products already made disappear. All cells, no matter what their origin, come to look alike and are indistinguishable from loose embryonic mesoderm (Hay, 1962). This is also apparent in the epidermis. For example, the gland cells of larval skin lose their special characteristics. This detooling for special projects spreads proximally until a blastema composed of detooled cells begins to retool for new projects. Schotté, Butler, and Hood (1941) demonstrated that the wave of dedifferentiation spreading proximally can be stopped prematurely on a young stump by grafting an already formed blastema to it.

It is just before appreciable amounts of new products appear in the blastema cells that the blastema begins to cut off further dedifferentiation behind it. The cells of a controlling blastema are already retooling for their new special duties. Such a blastema can be transplanted to a non-limb site and will continue to develop into a limb (Schotté, and Butler, 1941, 1944).

It would appear that when a portion of a limb no longer receives inhibitory messages from more distal regions it can transform to the missing distal regions. According to modern theory this would involve the use of no longer repressed DNA responsible for special RNA and special protein of a more distal region and also the repression of the special DNA to protein series which it was using in its more proximal position. Only when new special distal substances are being produced does the process of transformation to distal stop in more proximal regions. This system is such that when the entire distal-proximal series is complete there can be no further change.

For many years there was a debate over whether one kind of cell can transform to another. Then as the weight of the evidence in favor of transformation increased the question arose as to whether the cells of one territory could transform and function as part of another territory. This is an area of study in which theory sometimes overwhelmed experimental evidence. Late in the nineteenth century and through the early part of the twentieth the theory of differentiation by segregation of Moritz Nussbaum (1880) and August Weismann (1883, 1892) prevailed. There was reason to believe that at the one-cell stage of development all of the substances which would determine regional characteristics were present. It was further believed that, as the cells divided, these hypothetical organ-forming substances were divided and redivided many times until eventually all of the liver-forming, for example, were in one group and the lens-forming in another group. Differentiation was a process of segregation by cell division. One group of cells was exempt. They were the germ cells which would produce the next generation. They, the eggs and sperm, were thought to contain all the organ determiners for the next generation. Usually it was the egg that was believed to carry these special organ-forming substances. The fact of regeneration, especially in plants and some of the invertebrates where tiny pieces of one specialized region can transform to whole organisms, gave the proponents of the theory of differentiation by segregation a bit of trouble. The problem was resolved by inventing some elves. These were hypothetical germ-line cells which were supposed to operate throughout organisms. They retained all of the determiners. In some mysterious way, these intelligent cells knew when something was missing; they congregated and made the missing part. Much of the regeneration literature

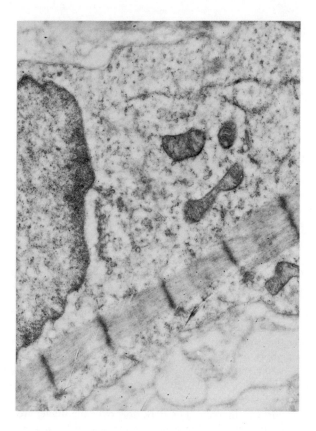

FIG. 3-5. AN ELECTRONMICROGRAPH OF A PORTION OF A CELL FROM A YOUNG BLASTEMA [*furnished by Elizabeth D. Hay*)].

This cell contains a myofibril running across the lower part of the figure at 30° to the base. The myofibril still present in this otherwise embryonic cell tells us that this cell had been part of a muscle.

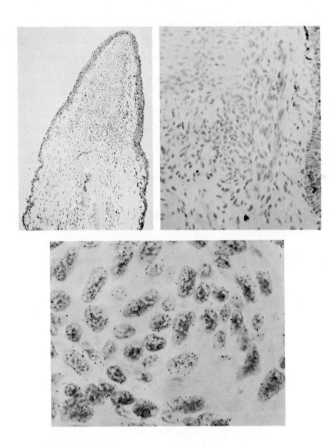

FIG. 3-6. TRANSFORMATION OF INTESTINAL BLASTEMA TO A LIMB CARTILAGE
           [*from the work of John Oberpriller*].

Intestinal blastema cells were first labeled with tritiated thymidine and then
implanted into limb blastemas shortly before the regeneration of the new skeleton.
Ten days later the limbs were prepared for histological study. At the upper left
is a low power photomicrograph of a stump capped by a cone-shaped blastema
in which the new skeletal pattern is visible. At the upper right is a higher-power
view of the lower right of the blastema, showing a whorl of new cartilage below
the center. In the lower photograph at still higher magnification the nuclei of the
cartilaginous whorl can be seen. The dark dots over the nuclei are silver grains
which had been activated by the electrons emanating from the tritium. The
tritium had been placed in intestinal cells and now the label appears in cartilage.
The cartilage cells are transformed cells of intestines, which never contain cartilage.
The possibility that the label had been transferred from intestinal to limb cells
was ruled out by Dr. Oberpriller's carefully controlled experiments.

of the last half century has been devoted to a consideration of this problem; we will go deeper into it in Chapters 6 and 7. The belief in totipotent elves in the limb was finally put to rest when Elizabeth Hay (1959) demonstrated with electron micrographs that some typical limb blastema cells could be shown to contain a telltale remnant of their former differentiated state. In Figure 3-5 a myofibril remains, indicating that this blastema cell was formerly part of a muscle fiber. At last it was no longer necessary to believe in reserve elves. Incidentally, the elves were much more viable than the theory they supported. The theory fell before Genetics, Cytology, and Developmental Biology many years earlier.

For many years, the only transformation in vertebrates acceptable to the general biological public was the transformation of iris to lens described in Chapter II. Here pigmented iris cells could be followed step by step as they lost their pigment and transformed to the new shapes of lens cells, at the same time becoming transparent. This was thought by many to be an exception.

Recently Florence Rose and Meryl Rose (1965) have demonstrated that epidermal cells tagged with a radioactive label can be followed step by step as they first become blastema cells and then cartilage, the first tissue to redifferentiate. Possibly some of the epidermal cells would have become parts of other tissues but the label was too dilute before the later-developing tissues such as muscle had appeared.

John Oberpriller (1967) has improved the method of checking for transformation. What he did was to take a part of a blastema just before it was to be determined in its original territory and implant it in a blastema of another territory. The host blastema was also about to differentiate. Since the cells were labeled just before they were to come under the influence of a new territory, they had to undergo only a very few cell divisions before differentiation. The label was still heavy after differentiation. By this method Oberpriller was able to trace tail cells and intestinal cells while they were developing into well integrated parts of limbs (Fig. 3-6).

THE ROLE OF NERVES

Whenever nerves are present in animals, both invertebrate and vertebrate, they are necessary for regeneration. It is a special role they play in regeneration because, in general, they are not needed during embryonic development. Limbs, for example, form without nerves as shown by Ross Harrison (1904) and by Viktor Hamburger (1928). What could nerves be needed for during regeneration of particular structures when the embryonic development of the same structures does not require them?

If nerves supplying a limb are cut, as reported by T. J. Todd in 1823, the limb cannot regenerate. Such a denervated limb is incapable of regenerating until the severed peripheral nerves regenerate and repopulate the limb. A denervated adult limb of a salamander responds to amputation like an X-rayed limb. The distal end of the stump does not lose structure rapidly. The wound heals normally as an epidermal sheet glides over the cut surface. Nothing like the normal extensive conversion to mesenchyme occurs (Fig. 3-7). When new tissue forms internally it is limited to scar formation and bone repair. The healing is usually excellent, but these denervated limbs, like X-rayed limbs and like limbs of the higher vertebrates, all respond to amputation in the same way. The stumps act as though nothing is missing. There seems to be a failure of communication —more about this later.

An experiment by Piera Locatelli (1929), began to shed light on the way in which nerves promote regeneration in salamanders. She cut the sciatic nerve in the hind limb, carefully extracted the proximal portion— still attached to the spinal cord—and led the cut end to the surface of the body close to the base of the limb. A regeneration bud arose directly over the deviated nerve ending and transformed to a limb. At first it was thought that limb nerves carry some special kind of limb-forming information. This proved to be wrong, as demonstrated by the school of Emile Guyénot (1927). Guyénot and Oscar Schotté (1926) and Daniel Bovet (1930) showed that if the same hind limb nerve, the sciatic nerve, was brought to the surface near a crest which extends along the middle of the back there was an outgrowth induced, but it became a piece of dorsal crest. The sciatic nerve when led to the surface farther back, at the base of the tail, caused an outgrowth, but it showed features of a tail. It is now quite clear that a deviated nerve can repolarize an area and cause tissue at its endings to behave like the most distal tissue in that area. It can then proceed to become a tail tip in a tail region or the distal end of a limb in the limb region. When the nerve is led to the surface at the base of a limb, the direction of communication must be changed. The tissue at the end of the nerve is no longer controlled by the tissue distal to it. Instead it behaves as if it itself were the most distal tissue. The base of a limb free from more distal influences and with a newly oriented communication system is set to produce everything not present distal to it—in this case, a whole limb.

When a nerve is deviated to the side of a limb part way along the length of the limb, a bud begins to form over the end of the nerve and that area

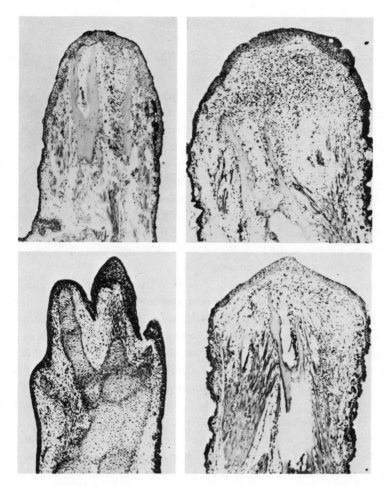

FIG. 3-7. NORMAL REGENERATION COMPARED WITH RELATIVELY LITTLE CHANGE IN X-RAYED LIMB AFTER AMPUTATION [*from S. M. Rose.* Jour. Gerontology, 22 (II): 28].

Upper left: A mid-longitudinal section of a stump of a forearm of the newt, *Triturus viridescens*, amputated 5 days before. There has been some loss of structure distally in the soft tissue, but the radius and ulna are intact and extend to the wound epithelium.

Upper right: A similar view 18 days after amputation shows loss of all internal tissue structure and its substitution by a blastema composed of relatively unorganized embryonic tissue.

Lower left: By 32 days after amputation, the new skeletal pattern can be seen emerging in the blastema.

Lower right: An X-rayed limb 31 days after amputation has healed with a wound epithelium. Its internal structural changes are much retarded and it appears only slightly ahead of the normal 5-day limb. Its bone still extends almost to the wound epithelium, and its tissues have not been converted to an appreciable blastema.

transforms not to a whole limb but just to the part that normally lies distal to the level at which the bud appears.

It is quite apparent that it is the information of a territory which is being used, as clearly pointed out by Guyénot (1927). The nerve is apparently necessary for the flow of information, and the direction of the nerve determines the direction in which the information will flow. In essence, what one does when deviating a nerve to the side of a limb is to short-circuit the flow of information. Instead of flowing down the length of a limb from distal to proximal, the controlling information starts from a new point at the end of the deviated nerve. The cells in that region are receiving information because nerves are present, but with the deviation away from the tip of the limb they behave as though that tip were missing and proceed to make the "missing" parts.

We have been considering the role of relatively small aggregations of neural tissue. These have not betrayed any morphogenetic effect of their own. Either they transmit information for other tissues, or they cause it to flow along other tissues. Spinal cords, much larger aggregations of neurons and connective tissue, are very effective morphogenetic agents in their own right. As shown by Sybil Holtzer (1956) a transplanted piece of larval spinal cord can act as a territory and direct the transformation of tissues around it. During embryonic and larval life it may be the chief molding agent (Howard Holtzer, 1959).

Several important things about the role of nerves in limb regeneration have been worked out by Marcus Singer and his associates. First of all there is a quantitative relationship involved. If the nerve supply in a salamander's limb is reduced to approximately a third of the number of fibers, regeneration of the limb will probably fail. With still fewer fibers it will certainly fail (Singer, 1946b). To return to the work of Locatelli, it was learned that all connections between peripheral nerves and the spinal cord could be cut and regeneration of a limb could proceed. It was not necessary to have intact spinal reflex arcs, but it was necessary to have the dorsal root ganglia and their sensory fibers intact. In fact regeneration could occur if only one very large dorsal root ganglion and its fibers remained. Locatelli suspected that this large aggregation of nerve-cell bodies was some special kind of regeneration-controlling organ. Later it was shown by Singer (rev. 1952) that there are no special regeneration-controlling nerves, but both sensory and motor nerves can function equally well in fostering limb regeneration. It is only because there are so many sensory nerves compared to motor nerves that there are enough sensory nerves to do the job after all motor nerves have been cut. Singer then went on to show that the sensory nerves can be removed and the motor nerves cut and allowed to regenerate back into the limb. As the motor nerves grew back in the absence of all or almost all sensory nerves they bifurcated more often and produced more than the normal number of motor fibers. There were enough of the regenerated motor fibers to foster limb regeneration in the absence of the sensory nerves (Singer, 1946a).

Singer (1959) now believes that the important number is the ratio between nerve fibers and amount of tissue to be controlled. It is only when the concentration of nerves relative to the other tissues is high that limb regeneration can occur. Possibly in the higher vertebrates and in the middle vertebrates where limb regeneration does not occur there are too few nerves.

It has been demonstrated by Singer with adult frogs (1954) and by Singer (1961) and by Simpson (1961) with lizards that one can induce fairly good, though never perfect, limb regeneration by deviating additional nerves to these normally non-regenerating limbs.

Now we return to the fact that in embryos nerves are not required for the original development of limbs. Strangely enough, as demonstrated by Chester Yntema (1959a, b), if nerves are prevented from entering a limb during its early development, it can then regenerate without the nerves. Furthermore, if nerves are later allowed to grow into these limbs which developed without nerves, and a part of the limb is amputated, it is very likely that two partially fused regenerates will arise. We know that all developing systems, both plant and animal, are polarized. It was clearly demonstrated by Ross Harrison for the amphibian limb bud as described above. Apparently, after a limb is innervated, polarization and the consequent control of the direction in which morphogenetic information will flow are taken over by the nerves. In these exceptional cases in which nerves were excluded for a long period and then allowed to enter, there seems to be a double set of controls. This is an area which is being explored further by Charles Thornton and his students. The aneurogenic limbs— those which never had nerves—have a tendency to form doubled structures during regeneration even if not repopulated by nerves (Thornton, in a lecture, 1967).

PATHWAYS OF CONTROL

The nerves seem to make it possible for regions where they end to transform to more distal parts. Long ago, clues concerning some special role of epidermis and nerves working in concert began to emerge from experiments.

The first observation leading to the understanding we now have was that if limb regeneration is to proceed contact between epidermis and some internal limb tissue or tissues must not be blocked. As long ago as 1906, Gustav Tornier, and again in 1921 Julius Schaxel demonstrated that if an amputation wound is covered by a seal of whole skin, regeneration does not begin. The next observations in the series were provided by Emil Godlewski (1928). Again with salamanders it was demonstrated that skin sewn over an amputation wound at the time of amputation prevents regeneration. However, when a day or so elapsed before the skin was sewn in place, regeneration could occur. M. E. Efimov (1933) showed why. If whole skin is not sewn over a limb during the first day after amputation or if a flap is left partially unsewed, epidermis can migrate into and over the wound surface. In the cases where limb skin is sewn over the surface after epidermis has migrated across the surface, the order of tissues from distal to proximal is epidermis of sewn skin flap, dermis of the flap, migrated epidermis, and immediately adjacent to the naked epidermis are the internal tissues of the limb. Many investigators have now confirmed that what gets limb regeneration started is the close contact of epidermis and internal tissues. When whole skin is applied to an amputation surface, the fibrous dermis acts as a barrier between epidermis and internal tissues.

The important internal tissue with which the epidermis must make contact is nerve. If a deviated nerve stimulates secondary limb formation, the nerve must be led close to the surface. Leon W. Polezhayev (1933), the great Russian student of regeneration, was aware of this, and he used to freshen a skin wound above a deviated nerve to stimulate regeneration.

Only later did it become clear that epidermis had to be free of dermis during early regeneration because an intimate contact between nerves and epidermis had to be established. Marcus Singer (1949) and Charles Taban (1949), after staining histological sections of limbs for nerves, learned that large bundles of nerve fibers enter the wound epithelium and ramify throughout it. Junctions between the nerve sprigs and the epidermal cells are very close and in electronmicrographs by Elizabeth Hay (1960) have the appearance of nerve synapses.

It is important that only after this contact is established do the tissues at the end of a stump begin to transform. Charles Thornton has studied the correlation between epidermis and nerve and has learned that wherever nerves and epidermis join, regeneration can occur. Thornton (1954, 1956) also thinks that failure to regenerate limbs by higher vertebrates may result from a failure to establish this contact.

Frog tadpoles can regenerate their hind limbs until metamorphosis begins. Then the animal is transforming from an aquatic to a terrestrial animal. There are many changes (Wilder, 1925). The skeleton becomes sturdier, and the skin becomes tougher and more resistant to drying. It also seals its surface wounds with tough patches of skin. A frog tadpole which can still regenerate its limbs has a skin like aquatic salamanders. After amputation there are several days when the migrated epidermal wound covering is not underlain by dermis. Nerve fibers are free to enter the epidermis and do so in large numbers. At the same time, epidermis piles up at the tip and an apical cap forms. Then there is transformation of distal stump tissues to a blastema.

More advanced tadpoles and metamorphosed frogs have acquired a better healing system which enables them to survive on land. Part of the change from an aquatic to a terrestrial habitat involved an increased blood pressure. The rapid closure of a wound and prevention of excess bleeding seems to have survival value on land. These terrestrial animals and the tadpoles about to become terrestrial have migration of epidermis as the initial step in wound healing. It is rapidly followed by a purse-string action which brings whole skin, including dermis, pinching in over the wound. At the same time, new dermal fibers form a mat beneath the epidermis. Nerve fibers do not enter the epidermis of these non-regenerating vertebrates. A healing feature valuable for survival may be a major reason for failure to regenerate.

Thornton has made other observations along this line. Both larval and adult salamanders fail to regenerate limbs after denervation. After nerves have regenerated and repopulated limbs some larval limbs regenerate, others fail to regenerate as do all repopulated adult limbs. The adult limbs which fail have formed an appreciable dermal barrier between epidermis and approaching nerve fibers by the time they return to the stump. There is also a dermal barrier in those larval limbs which fail to regenerate after repopulation by nerves. Those larval limbs which do regenerate did not yet have a barrier to the returning nerve fibers. All one has to do to establish conditions for regeneration in those with a dermal barrier is to cut it away along with the epidermis. Epidermis migrates in from the periphery of the wound, nerves invade it, and limb regeneration ensues (in discussion at end of Singer, 1959).

Why is this contact so important? Without it transformation to more distal including the initial step of dedifferentiation scarcely gets underway. We do not know from work on Amphibia what the nature is of the information which causes the transformation. It seems to come to the cells because a contact has been made. The major pathway as indicated above seems to be from the stump, distally through the epidermis and then proximally through the internal tissues. Without a flow of information, very limited changes follow amputation. These are confined to wound healing and scar formation.

Although the major arc of control seems to be the one passing through epidermis, this cannot be the only arc. Ordinarily this arc may be the most important one because the major nerve supply is afferent and much of it contacts epidermis. In addition, the distal position of the epidermis may be important as noted above. When the nerve supply is severed and the return of the afferent or sensory supply is blocked, the efferent or motor fibers may return and branch more than usually and may then provide a sufficient amount of nerve tissue to promote regeneration (Singer, 1946a). These motor fibers do not contact epidermis (Sidman and Singer, 1960; Thornton, 1960). Instead, they make their connections with internal tissues. In this case it seems unlikely that morphogenetic control is by way of the epidermis. Instead it seems likely that other tissue arcs are passing the controlling information.

Although regeneration studies on Amphibia have provided valuable information, most of the modern studies dealing with the general problems of regeneration are done with worms, hydroids, plants, and Protozoa. The reason for this is that studies have advanced far enough along all lines to indicate that there are general principles. The way to learn them in the shortest time is to experiment with rapidly regenerating organisms.

However, before we leave the Amphibia, there is one bit of modern work which should be covered. This is a first in its field. Robert Becker (1961), like Alberto Monroy (1941) before him, has shown that there are electrical potential differences along limbs. Becker has studied changes in potential differences during regeneration in limbs of regenerating salamanders and non-regenerating frogs. What he did was to study how the charge at the tip of the limb varies with respect to the base during regeneration. As can be seen from Figure 3-8, tips of limbs of both salamanders and frogs are slightly negative. After amputation, the distal ends of both salamander's and frogs' limbs become highly positive. In the salamander, shortly after amputation, the tip of the stump begins to become more negative. The peak of negativity is reached just over two weeks after amputation. During the period of highest negativity dedifferentiation and transformation are proceeding rapidly. By the time all new structures have formed but before they have achieved full size, the potential differences along the limb have returned to normal.

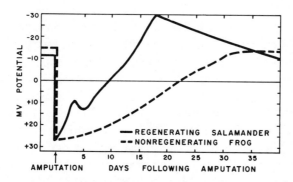

FIG. 3-8.   COMPARISON OF CHANGES IN ELECTRICAL POTENTIAL IN THE REGENER-
ATING SALAMANDER AND THE NON-REGENERATING FROG LIMB [*furnished*
by R. O. Becker].

The curve is quite different in the frog's limb. The tip remains electropositive for a long period and no regeneration occurs.

There are two questions here. Is the electronegativity at the tip important for regeneration? This is not yet known, but Becker has shown that
continued electronegativity is dependent upon the presence of nerves. The
potential differences along the limb decrease after nerves are severed.
Possibly the nerves are important to regeneration because they create
potential differences. The second question is, "how could potential
differences be important?" Recent studies on lower organisms—detailed
in subsequent chapters—indicate that regional regeneration is controlled
by the movement of charged particles in a bioelectric field. Becker (1967)
has demonstrated that extremely weak fields of the order of magnitude of
bioelectric fields can be applied to enucleated amphibian red cells *in vitro*
and cause them to lose their specialized form. This he thinks of as dedifferentiation. It should be recalled that in the salamander's regenerating
stump the end becomes negative, and this is the time when dedifferentiation
begins. There is extremely little dedifferentiation in the non-regenerating
frog's limb. Correlated with this is a failure of the end to become negative.

Stephen Smith (1967) has been able to stimulate dedifferentiation and
limited regeneration by applying weak electric fields to the ends of limb
stumps of frogs. This was accomplished by implantation of bimetallic
couples. Short pieces of fused platinum and silver were implanted close to
the distal end of the stump. These couples generated 0.2 volts in amphibian
Ringer's solution. Regeneration occurred when either the Pt+ was at the
distal making the distal end (+) or when the end was made (−) by having

the Ag in the distal position. The fact that regeneration began when either pole was distal will have to be explained. The effect was apparently an effect of a created field rather than simple irritation by the metal, because neither platinum nor silver alone produced a comparable effect. Couples prevented dermis from forming at the tip under the epidermis in 11 of 15 cases, whereas it did form in 19 of 20 cases when the single metals were inserted.

Hypotheses based on information covered in both the early and later parts of this chapter have led several young workers in this field to new experiments the results of which are just becoming available.

Jean Oberpriller (1968) is trying to answer the question of the nature of the suppressive interaction when connected limbs regenerate at an angle to each other. She knew that X-irradiation blocks regeneration and that it seems to do this by blocking the transmission or transportation of morphogenetic control. Could it be that two control systems meeting at an angle interfere with each other? She reasoned that if this were so, regeneration of one of two limbs grafted together at an angle should be perfect if the other limb was X-rayed. This turned out to be so. Her control normal limbs grafted together at an angle do interfere with each other, just as Monroy and Goss had demonstrated. If one member of the angled pair is X-rayed the other may regenerate perfectly (Figs. 3-4 and 3-9). These findings leave us with the knowledge that there is some kind of communication system operating during regeneration. Furthermore, the nature of it is such that each of two contiguous non-parallel systems can neutralize the effect of the other.

When one considers this new finding in conjunction with other new information, the theory of polarized control is strengthened. As noted above, both X-irradiation and denervation cause regeneration to fail in the same way. There is relatively little change in limb stumps after these treatments. The remaining parts do not begin to transform to the missing distal parts.

The work of Becker makes one think that nerves are important to regeneration, because their activity yields a polarized electrical field. If denervation and X-irradiation both block regeneration in the same way, the normal sequence of changes in electrical potential during regeneration should not be found after either denervation or X-irradiation. Jean Oberpriller is studying this now.

Another point which has been stressed above is that nerves must make contact with other tissues if the information controlling regeneration is to flow. Under normal conditions, the greatest area of contact is between nerves and epidermis. This contact is not established in X-rayed limbs.

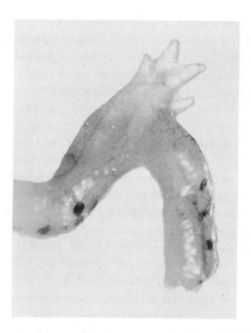

FIG. 3-9.   LINEAR MORPHOGENETIC CONTROL BLOCKED BY X-IRRADIATION [*from Jean Oberpriller*, Jour. Exptl. Zool. 168: *403, 1968*].

When two normal limbs are joined as above, no regeneration or single-spike regeneration occurs (Fig. 3-4). In this case, the limb on the right was X-rayed. A perfect hand aligned with the left limb had regenerated. X-rayed tissues do not influence the regeneration process proceeding distal to them.

If normal epidermis is added to X-rayed limbs, nerves and epidermis do make contact (Rose and Rose, 1967). Normal regeneration involving both the normal and the X-rayed tissues occurs. Apparently the polarized message transport system is restored. It now becomes important to know whether the establishment of the contacts between nerves and other tissues restores a polarized electrical system. This is to be studied.

There is now additional information that the skin is part of a polarized control system. It should be recalled that Umanski and his co-workers had demonstrated that X-rayed limbs could regenerate when given grafts of normal skin properly oriented or when turned through 180°. There was no recovery if the limbs received skin grafted at 90° to its normal axis. It has been demonstrated that charged particles can be transported through skin more easily along its longitudinal axis than in other directions. Harry

Settles (1969) started with the above information and several other bits. As was known, either the skin or the rest of the limb without its skin can control regeneration distal to it. If regeneration does require polarized fields and if such fields are generated in skin and the other tissues when they are in intimate contact with nerves, the turning of skin 90° on a normal stump should set up two fields. Harry Settles (1969) has looked for evidence of this by simply cutting off pieces of skin and replacing them at 90° to their original orientation. One would expect such crossed fields to interfere and no regeneration to occur, or, if they were sufficiently isolated, two hands to appear. Most of the limbs regenerated after the operation. More than 30 percent of the regenerates were two-handed. This indicates that there is a polarized controlling system both in skin and in non-skin.

There is also new evidence concerning transformation from one cell type to another during regeneration. Donald Donaldson (1969) has followed epidermis of the tail of salamanders during tail regeneration and has found that some remains epidermis while some becomes tip of the regenerating spinal cord and cartilage. The transformation of epidermis to non-epidermal tissues has been repeatedly questioned over the years. The doubts arise from the fact that label might be transferred from one cell type to another or that another type of cell such as a labeled blood cell might enter the limb and transform. John Oberpriller (1967), while following the transformation of intestinal cells, had shown that a labeled and killed graft did give up some of its label to regenerating limb cells: the amount was of the order of one percent of the original label, an insignificant percentage compared to the 10 percent or more of the original label which has been observed in the differentiating cells. Trygve Steen (1968) has followed doubly-labeled cartilage cells. They were triploid, sometimes showing 3 nucleoli, and were in addition labeled with tritiated thymidine. In testing for how much radioactive label might be transferred from cell to cell, he grafted some triploid cells not labeled with tritiated thymidine into limbs of animals which had previously received tritiated thymidine. As many as 5 silver grains could be seen over some triploid cells which should have shown no label. As Steen suggested, this would make it impossible to use counts in this range as evidence for transformation. This is true, but in the Oberpriller and Donaldson cases the counts observed over the differentiated cells were much greater than 5, in fact more than 5 times as great.

In the Rose and Rose 1965 work, sometimes cartilage cells with grain counts as low as 5 were thought to have arisen from former epidermal cells which had been labeled. However, the cells of the preceding generations in the cartilage region had 10 to 20 grain counts. One can follow epidermal cells as their grain count decreases by approximately half with each cell division. By the time some of these have become mesenchymal, their grain counts average in the 15–30 count range, appreciably higher than one could expect from transfer.

Another doubt concerning the tritiated thymidine method of labeling is that there might be a circulating pool of tritiated thymidine. For example, tritiated thymidine might remain in the bloodstream for days. If this were so, cells other than those which were followed might be labeled. Confusion would result because not only the descendants of the original cells of one type which had been labeled but others as well would be labeled.

Another possibility is that tritiated thymidine might be picked up by cells somewhere else in the body, held for a time, and then released later. If the released tritiated thymidine were then picked up by regenerating limb cells of non-epidermal origin, two types of cells would be labeled and they would not be distinguishable. To check this possibility, Donaldson injected newts with tritiated thymidine. Several days later, lenses were removed. If there had been labeled thymidine available, the transforming iris cells would have been labeled. They were not. This indicates that there is no long-lived circulating pool which might label cells other than those one intended to trace.

Even though label cannot come from afar, it was still possible that it might be transferred from labeled cells to nearby unlabeled cells. This has been tested by taking a patch of skin from a newt and labeling the newt after the patch had been removed. Four hours later the unlabeled skin was put back in its original site on the animal. It remained unlabeled for days and weeks. This means that there is neither sufficient tritiated label remaining after the first four hours nor is there transfer from nearby cells.

Another possibility which has been considered by the investigators is that developing blood cells labeled in their sites of formation might mature and enter limbs or tails where they might transform to cells of an appendage. One can find labeled blood cells outside blood vessels and intimately mixed with the resident cells of the appendage. Donaldson has made a quantitative study which indicates that when epidermis and blood are both labeled, there is not enough label in the blood cells which remain in a tail at the time of blastema formation to account for more than a small fraction of the label in regenerating cartilage and spinal cord. The only tissue which contains enough is the epidermis.

After these carefully controlled studies it is now quite certain that cells of appendages do transform from one type to another during the course of regeneration.

References Cited

Balinsky, B. I. 1925. Transplantation des Ohrbläschens bei Triton. *Arch. Entw-mech. Org.* 105:718–731.

———. 1926. Weiteres zur Frage der experimentallen Induktion einer Induktion einer Extremitätenanlage. *Arch. Entw-mech. Org.* 107:679–683.

———. 1927. Uber experimentelle Induktion der Extremitätananlage bei Triton mit besonderer Berücksichtigung der Innervation und Symmetrieverhältnisse derselben. *Arch. Entw-mech. Org.* 110:71–88.

Becker, R. O. 1961a. Search for evidence of axial current flow in peripheral nerves of salamanders. *Science* 134:101–102.

———. 1961b. The bioelectric factors in amphibian limb regeneration, *J. Bone Joint Surg.* 43A: 643–656.

———. 1967. The electrical control of growth processes. *Medical Times* 95:657–669.

Bischler, V. 1926. L'influence du squelette dans la régénération et les potentialités des divèrs territoires du membre chez *Triton cristatus. Rev. Suisse Zool.* 33:431–560.

Bovet, D. 1930. Les territoires de régénération; leurs propriétés étudiées par la méthode de déviation du nerf. *Rev. Suisse Zool.* 37:83–146.

Breedis, C. 1952. Induction of accessory limbs and of sarcoma in the newt (*Triturus viridescens*) with carcinogenic substances. *Cancer Res.* 12:861–866.

Brunst, V. V. and Figge, F. H. J. 1951. The development of secondary tails in young axolotls after local X-ray irradiation. *J. Morphol.* 89:111–133.

Brunst, V. V. and Scheremetjewa, E. A. 1937. Ist es möglich, die Regenerationsfähigkeit einer *Triton* Extremität zu vernichten, ohne ihre Lebensfähigkeit zu schädigen? *Byul. Eksperim. Biol. i. Med. 3*:397–399.

Butler, E. 1955. Regeneration of the urodele limb after reversal of its proximo-distal axis. *J. Morphol.* 96:265–282.

Butler, E. G. and Blum, H. F. 1955. Regenerative growth in the urodele fore-limb following ultraviolet radiation, *J. Natl. Cancer Inst.* 15:877–889.

Chen, L. 1933. Regeneration von kleinen Knochenstücken (Dritteln und Sechsteln) im Innern von Molchextremitäten. *Zool. Jahrb. Abt. Allgem. Zool. Physiol. Tiere* 53:151–172.

Della Valle, P. 1913. La doppia rigenerazione inversa nelle fratture della zampe di *Triton. Boll. Soc. Nat. Napoli* 25:95–161.

Dent, J. N. 1954. A study of regenerates emanating from limb transplants with reversed proximo-distal polarity in the adult newt. *Anat. Rec.* 118:841–856.

Donaldson, D. J. 1968. An autoradiographic study of the origin of the tail blastema in the newt, *Triturus viridescens*. Doctoral Dissertation, Tulane University Library.

Efimov, M. E. 1933. Uber den Mechanismus der Regenerationsprozesse III. Bleibt die Polarität der Extremität beim Regenerationsprozess erhalten und die Rolle der inneren Teile des Organs in diesem Prozess. *Biol. Zhur.* 2: 220–231 [German Summary].

Faber, J. 1960. An experimental analysis of regional organization in the regenerating forelimb of the axolotl (*Amblystoma mexicanum*). *Arch. Biol.* Liége 71: 1–72.

Farinella-Ferruzza, N. 1957. Transformazione di coda in arto nei trapianti xenoplastici di bottone codale di Triton cristatus su Discoglossus pictus. *Acta Embryol. Morphol. Exptl.* 1: 171–187.

Filatow, D. 1927. Aktivierung des Mesenchyms durch eine Ohrblase und einen Fremdkörper bei Amphibien. *Arch. Entw-mech. Org.* 110: 1–32.

Flat-Hauser, E. and Przibram, H. 1930. Regeneration der langen Knocken nach teilweiser Entfernung im Innern der Molchextremitäten (Triton cristatus Laur). *Arch. Entw-mech. Org.* 122: 237–250.

Fritsch, C. 1911. Experimentelle Studien über Regenerationsvorgänge des Gliedmassenskeletts der Amphibien. *Zool. Jahrb. Abt. Allgem. Zool. Physiol. Tiere* 30: 377–472.

Glade, R. W. 1957. The effects of tail tissue on limb regeneration in *Triturus viridescens*. *J. Morphol.* 101: 477–522.

———. 1963. Effects of tail skin, epidermis, and dermis on limb regeneration in *Triturus viridescens* and *Siredon mexicanum*. *J. Exptl. Zool.* 152: 169–194.

Godlewski, E. 1928. Untersuchungen über Auslösung und Hemmung der Regeneration beim Axolotl. *Arch. Entw-mech. Org.* 114: 108–143.

Goss, R. J. 1956a. The relation of bone to the histogenesis of cartilage in regenerating forelimbs and tails of adult *Triturus viridescens*. *J. Morphol.* 98: 89–123.

———. 1956b. The unification of regenerates from symmetrically duplicated forelimbs. *J. Exptl. Zool.* 133: 191–209.

———. 1957. The relation of skin to defect regulation in regenerating half limbs. *J. Morphl.* 100: 547–563.

Graper, L. 1922. Extremitätentransplantationen an Anuren II. Mitteilung: Reverse Transplantionen. *Arch. Entw-mech. Org.* 51: 587–609.

Guyénot, E. 1927. Le problème morphogénétique dans la régénération des Urodèles determination et potentialité des régénérats. *Rev. Suisse Zool.* 34: 127–154.

Guyénot, E. and Schotté, O. 1926. Demonstration de l'existence de territoires spécifiques de régénération par la méthode de la deviation des troncs nerveux. *Compt. Rend. Soc. Biol.* 94: 1050–1052.

Hamburger, V. 1928. Die Entwicklung experimentell erzeugter nervenloser und schwach innervierter Extremitäten von Anuren. *Arch. Entw-mech. Org.* 114:272–363.

Harrison, R. G. 1904. An experimental study of the relation of the nervous system to the developing musculature in the embryo of the frog. *Am. Jour. Anat.* 3:197–220.

———. 1918. Experiments in the development of the forelimb of *Amblystoma*, a self-differentiating, equipotential system. *J. Exptl. Zool.* 25:413–462.

———. 1921. On relations of symmetry in transplanted limbs. *J. Exptl. Zool.* 32:1–136.

Hay, E. D. 1958. The fine structure of blastema cells and differentiating cartilage in regenerating limbs of *Amblystoma* larvae. *J. Biophys. Biochem. Cytol.* 4:538–592.

———. 1959. Electronmicroscopic observations of muscle dedifferentiation in regenerating *Amblystoma* limbs. *Develop. Biol.* 1:555–585.

———. 1960. The fine structure of nerves in the epidermis of regenerating salamander limbs. *Exptl. Cell Res.* 19:299–317.

———. 1962. Cytological studies of dedifferentiation and differentiation in regenerating amphibian limbs. In *Regeneration*. ed. D. Rudnick, pp. 177–210. New York: Ronald Press.

Holtzer, H. 1959. The development of mesodermal axial structures in regeneration and embryogenesis. In *Regeneration in Vertebrates*, C. S. Thornton, ed. pp. 15–33. Chicago Ill.: University of Chicago Press.

Holtzer, S. 1956. The inductive activity of the spinal cord in urodele tail regeneration. *J. Morphol.* 99:1–40.

Kasanzeff, W. 1930. Histologische Untersuchungen über die Regenerationsvorgänge beim Anlegen von Ligaturen an die Extremitäten beim Axolotl. *Arch. Entw-mech. Org.* 121:658–707.

Kiortsis, V. 1953. Potentialités du territoire patte chez le Triton. *Rev. Suisse Zool.* 60:301–410.

Kurz, Oskar. 1922. Versuche über Polaritätsumkehr am Tritonenbein. *Arch. Entw-mech. Org.* 50:186–191.

Lazard, L. 1959. Influence des greffes homologues et hétérologues sur la morphologie des régénérats de membres chez *Amblystoma punctatum*. *Compt. Rend.* 249:468–469.

Locatelli, P. 1929. Der Einfluss des Nervensystems auf die Regeneration. *Arch. Entw-mech. Org.* 114:686–770.

Luther, W. 1948. Zur Frage des Determinationzustandes von Regenerationsblastemen. *Naturwissenschaften* 35:30–31.

Milojevic, B. D. 1924. Beiträge zur Frage über die Determination der Regenerate. *Arch. Entw-mech. Org.* 103:80–94.

Monroy, A. 1941. Ricerche sulle correnti elettriche derivabili dalla superficie del corpo di Tritoni adulti normali e durante la rigenerazione degli arti e della coda. *Pubbl. staz. zool. Napoli* 18:265–281.

————. 1946. Ricerche sulla rigenerazione degli Anfibi urodeli. Nota III. Osservazioni su rigenerati formati si su doppie superfici de sezione e considerazione sui processi determinativi della rigenerazione. *Arch. Zool. Ital.* 31:151–172.

Nassonov, N. V. 1930. Die Regeneration der Axolotlextremitäten nach Ligaturanlegung. *Arch. Entw-mech. Org.* 121:639–657.

Nicholas, J. S. 1924. The response of the developing limb of *Amblystoma punctatum* to variations in orientation of the surrounding tissue. *Anat. Rec.* 29:108.

Nussbaum, M. 1880. Zur Differenzierung des Geschlechts im Tierreich. *Arch. mikr. Anat. Entwicklungsmech.* 18:1–121.

Oberheim, K. W. and Luther, W. 1958. Versuche über die Extremitätenregeneration von Salamanderlarven bei umgekehrter Polarität des Amputationsstumpfes. *Arch. Entw-mech. Org.* 150:373–382.

Oberpriller, Jean C. 1968. The action of X-irradiation on the regeneration field of the adult newt, *Diemictylus viridescens*. *J. Exptl. Zool.* 168:403–421.

Oberpriller, John. 1967. A radioautographic analysis of the potency of blastemal cells in the adult newt, *Diemictylus viridescens*. *Growth* 31:251–296.

Perri, T. 1956. Induzione di arti sopranumerari negli Anfibi (particolarmenti negli Anuri) mediante sostanze citolotiche. *Experentia* 12:125–135.

Polejaiev, L. 1936. La valeur de la structure de l'organe et les capacités du blastème régénératif dans le processus de la détermination du régénérat. *Bull. Biol.* France Belg. 70:54–85.

Polezhayev, L. W. 1933. Uber Resorption und Proliferation sowie über die Verhältnisse der Gewebe zu einander bei der Renegeration der Extremitäten des Axolotls. *Zeit. f. Biol.* 2:368–386 [Russ. with German Summary].

Przibram, H. 1931. *Connecting Laws in Animal Morphology* London: University of London Press.

Rose, F. C. and Rose, S. M. 1965. The role of normal epidermis in recovery of regenerative ability in X-rayed limbs of *Triturus*. *Growth* 29:361–393.

Rose, F. C. and Rose, S. M. 1967. Nerve penetration and regeneration after X-rayed limbs of *Triturus viridescens* were covered with unX-rayed epidermis. *Growth* 31:375–380.

Rose, S. M. 1964. Regeneration, Chapt. 10, In *Physiology of the Amphibia*, (J. Moore, ed.) pp. 545–622). New York: Academic Press.

Ruben, L. N. 1960. An immunobiological model of implant-induced urodele supernumerary limb formation. *Amer. Nat.* 94:427–434.

Schaxel, J. 1921. Auffassunger und Erscheinungen der Regeneration. *Arb. a. d. Geb. der Exp. Biol.* 1:1–99.

Schotté, O. E. 1940. The origin and morphogenetic potencies of regenerates. *Growth* 4 (Suppl. 1): 59–76.

Schotté, O. E. and Butler, E. G. 1941. Morphological effects of denervation and amputation of limbs in urodele larvae. *J. Exptl. Zool.* 87:279–322.

———. 1944. Phases in regeneration of the urodele limb and their dependence upon the nervous system. *J. Exptl. Zool.* 97:95–121.

Schotté, O. E., Butler, E. G., and Hood R. T. 1941. Effects of transplanted blastemas on amputated nerveless limbs of urodele larvae. *Proc. Soc. Exptl. Biol. Med.* 48:500–503.

Schwidefsky, G. 1935. Entwicklung und Determination der Extremitäten-regenerate bei den Molchen. *Arch. Entw-mech. Org.* 132:57–114.

Settles, H. E. 1967. Supernumerary regeneration caused by ninety degree rotation in the adult newt, *Triturus viridescens.* Doctoral Dissertation, Tulane University Library.

Sidman, R. L. and Singer, M. 1960. Limb regeneration without innervation of the apical epidermis in the adult newt, *Triturus. J. Exptl. Zool.* 144:105–109.

Simpson, S. B., Jr. 1961. Induction of limb regeneration in the lizard, *Lygosoma laterale,* by augmentation of the nerve supply. *Proc. Soc. Exptl. Biol. Med.* 107:108–111.

Singer, M. 1946a. The nervous system and regeneration of the forelimb of adult *Triturus.* IV. The stimulating action of a regenerated motor supply. *J. Exptl. Zool.* 101:221–240.

———. 1946b. The nervous system and regeneration of the forelimb of adult *Triturus.* V. The influence of number of nerve fibers, including a quantitative study of limb innervation. *J. Exptl. Zool.* 101:299–337.

———. 1949. The invasion of the epidermis of the regenerating forelimb of the urodele, Triturus, by nerve fibers. *J. Exptl. Zool.* 111:189–210.

———. 1954. Induction of regeneration of the forelimb of the postmetamophic frog by augmentation of the nerve supply. *J. Exptl. Zool.* 126:419–471.

———. 1959. The influence of nerves on regeneration, in *Regeneration in Vertebrates* (C. S. Thornton, ed.) pp. 59–80 Chicago: Univ. Chicago Press.

———. 1961. Induction of regeneration of body parts in the lizard, *Anolis. Proc. Soc. Exptl. Biol. Med.* 107:106–108.

Singer, M., Rzehak, K., and Maier, C. S. 1967. The relation between caliber of the axon and the trophic activity of nerves in limb regeneration. *J. Exptl. Zool.* 166:89–98.

Smith, S. D. 1967. Induction of partial limb regeneration in *Rana pipiens* by galvanic stimulation. *Anat. Rec.* 158:89–97.

Spallanzani, Abbé. 1769. *An essay on Annual Reproductions,* translated from the Italian, 1768, by M. Maty, London.

Steen, T. 1968. Stability of chondrocyte differentiation and contribution of muscle to cartilage during limb regeneration in the axolotl (*Siredon mexicanum*). *J. Exptl. Zool.* 167:49–78.

Taban, C. 1949. Les fibres nerveuses et l'épithélium dans l'édification des régénérats de pattes (in situ ou induites) chez le triton. *Arch. Sci.* (Geneva) 2: 553.

Thornton, C. S. 1942. Studies on the origin of the regeneration blastema in *Triturus viridescens. J. Exptl. Zool.* 89: 375–390.

――――. 1954. The relation of epidermal innervation to limb regeneration in *Amblystoma* larvae. *J. Exptl. Zool.* 127: 577–601.

――――. 1956. Epidermal modifications in regenerating and non-regenerating limbs of anuran larvae. *J. Exptl. Zool.* 131: 373–394.

――――. 1960. Regeneration of asensory limbs of *Amblystoma* larvae. *Copeia.* pp. 371–373.

――――. 1967. [Lecture].

Todd, T. J. 1823. On the process of reproduction of the members of the aquatic salamander. *Quart. J. Sci. Arts. Lib.* 16: 84–86.

Tornier, G. 1906. Kampf der Gewebe im Regenerat bei Begünstigung der Hautregeneration. *Arch. Entw-mech. Org.* 22: 348–369.

Trampusch, H. A. L. 1959. The effect of X-rays on regenerative capacity. In *Regeneration in Vertebrates* (C. S. Thornton, ed.) pp. 83–99. Chicago, Illinois: Univ. Chicago Press.

Umanski, E. 1937. Untersuchungen des Regenerationsvorganges bei Amphibien mittels Auschaltung der einzelnen Gewebe durch Röntgenbestrahlung. *Biol. Zhur.* 6: 757–758 [Russian-German summary].

――――. 1938a. The regeneration potencies of axolotl skin studied by means of exclusion of the regeneration capacity of tissues through exposure to X-rays. *Byul. Eksperim. Biol. i. Med.* 6: 141–145.

――――. 1938b. A study of regeneration of the axolotl limb upon substitution of the inner tissues by dorsal musculature. *Byul. Exsperim. Biol. i. Med.* 6: 383–386.

Umansky, E. E., Tkatsch, V. K., and Koudokotsev, V. P. 1951. Anisotropic diélectrique de la peau chez l'Axolotl. *Compt. Rend. Acad. Sci.* URSS 76: 465–467 (rev. by Kiortsis, 1953).

Waddington, C. H. 1938. Regulation of amphibian gastrulae with added ectoderm. *J. Exptl. Biol.* 15: 377–381.

Weismann, A. 1883. Die Entstehung der Sexualzellen bei den Hydromedusen Gustav Fischer, Jena. 295 pp.

――――. 1892. *The germ plasm.* English translation by W. N. Parker and H. Ronnfeldt. New York: Scribner.

Weiss, P. 1924. Regeneration aus doppeltem Extremitätanquerschnitt (an *Triton cristatus*). *Anz. Akad. Wiss. Wien Math. Naturw. Kl.* 61: 45–46.

――――. 1925. Unabhängigkeit der Extremitätenregeneration vom Skelett. *Arch. Mikr. Anat. Entwicklungsmech.* 104: 359–394.

――――. 1926. Ganzregenerate aus halbem Extremiteätnquerschnitt. *Arch. Entw-mech. Org.* 107: 1–53.

Wilder, I. W. 1925. *The Morphology of Amphibian Metamorphosis.* Smith College 50th-Anniversary Publication, Northampton, Mass.

Yntema, C. 1959a. Regeneration in sparsely innervated and aneurogenic fore-limbs of *Amblystoma* larvae. *J. Exptl. Zool.* 140: 101–124.

———. 1959b. Blastema formation in sparsely innervated and aneurogenic fore-limbs of *Amblystoma* larvae. *J. Exptl. Zool.* 142: 423–439.

CHAPTER FOUR

The Transport of Messages Controlling Growth and Differentiation in Plants

Most of the important developmental relationships which we now know to be general were first studied in plants.

The modern generalization—that any one region uses part of its genome and retains the rest in an unused state—is based on a nineteenth-century botanical study. The knowledge that whole plants can arise from detached parts of plants must go back to the early days of plant husbandry, Willow posts, for example, had a way of becoming trees. It was Fritz Regel (1876) who demonstrated with microscopic sections that differentiated cells of *Begonia* leaves could transform to whole plants. One epidermal cell inside its cellulose wall would start to divide. A few adjacent cells also still enclosed began to divide. In time the old walls were broken through, and a bud of growing embryonic tissue bulged out. This meristem arising from cells performing a limited function could then go on and produce a whole plant. Here was a very neat demonstration that a few differentiated cells of one special tissue contained all the information needed for a whole plant.

Even before botanists of the nineteenth century were beginning to learn about the totipotency of differentiated cells, another relationship to be of value in understanding differentiation was being learned. It arose from

79

the oft-repeated observation by farmers that some crops cannot be grown a second year in the same field without a drop in production. Yields were higher when the crops were rotated. Augustin-Pyramus de Candolle discussed this in his Physiologie Vegetale III (1832). He thought the plants added something to the soil which retarded both their germination and their growth, and that a species could adversely affect members of the same species, genus, and even members of the same family. In addition he knew that one should not try to replace old shade trees with young of the same kind.

Only one experiment was mentioned by de Candolle. He had seen M. Macaire demonstrate that kidney beans do not develop in water which had bathed the roots of kidney bean plants, but that wheat could grow very well in the same water. This was the beginning of the knowledge that like inhibits like. It has been demonstrated in a great variety of organisms since then (rev. by Rose, 1960; Rose and Rose, 1961, 1965). In addition, the phenomenon of like inhibiting potentially like parts is known for many organisms. Details are given throughout the book.

Another early experiment with plants led to further work which is important to our understanding of differentiation and growth control. Charles Darwin and his son Francis (1880) were trying to learn how plants turn and grow toward light. With a darkened room, lamps, some growing plants, and simple covers for shading parts of the plants, they learned something important. They discovered that there is a region near the end of a growing shoot which is sensitive to light. If this region is lighted from one side, it causes a region below it, which itself needs to receive no light, to grow less than the same level on the unlighted side. Two things of importance were learned from these simple experiments of the Darwins. First, a region can send messages that will control growth in another region. Second, the message was transmitted or transported along the main axis in linear fashion resulting in an asymmetrical growth response.

Another beginning was made by Hermann Vöchting (1885). He isolated pieces from the liverwort, *Marchantia*. Whether the isolates were large or small, the rule was that only one part of each transformed to a new plant. It was the most anterior region closest to the original midline which could transform. This was always true of isolates from young plants. From middle-aged isolates two regenerates sometimes arose. In older isolates the incidence of regeneration of two plants increased, and sometimes even three formed. This was the beginning of the study of regional, polarized dominance. The fact is noted that dominance tended to decrease with age.

The last early-discovered relationship to be considered is related to the dominance studies of Vöchting. He and other botanists had learned from isolating pieces of plants that there was a greater tendency to regenerate new shoots from one end of an isolate and a tendency grading off in the opposite direction to regenerate roots. It had been learned from isolation of aerial parts of many plants in a moist environment that there were a number of potential tops of plants in the form of dormant buds along their lengths. When a growing top was not present, one or more potential tops could grow. Roots could also form. There were no already formed root buds. These had to form anew after isolation by transformation of twig or trunk regions. The great synthesis was made by William Burnett McCallum (1905). In his words:

"The plant possesses innumerable growing points either organized or potential, the vast majority of which must not be allowed to develop if the plant body is to retain anything like a definite organization. In most cases this development does not occur in the ordinary life of the plant, because these cells, capable of producing new organs, are held in check by those parts already growing. This non-development does not seem to be due to any lack of those conditions that favor growth, as nutrition and moisture; or to such influences as light and gravity; or to a lack of definite 'formative substance' but to some influence independent of all of these, which an organ, acting perhaps along the protoplasmic connections, is able to exert over other parts and so prevent their growth. When this influence is removed, the favorable growth condition, present all the time, permits the growth of the part to occur. In such a controlling influence of growing organs over the numerous potential growing points throughout the plant there exists very evidently a principle of fundamental importance in plant organization."

Much work followed these early studies. Interest spread to other organisms both plant and animal and general rules were discerned. This book is in a sense a synopsis of the science of polarized control of growth and morphogenesis which began with the several botanical studies of the nineteenth century outlined above.

Let us return to the study of cellular potency in plants. By 1950 when Clarence Sterling reviewed this field, all kinds of differentiated plant cells had been shown capable of transforming to whole plants or large parts of plants. In recent years F. C. Steward and his associates started with single differentiated cells of carrots, cultured them and obtained whole carrot plants which grew up and produced flowers and seed (Steward, Mapes, and Mears, 1958; Steward, 1961).

The other lines of study which began in the nineteenth century have fused, and today we have a fairly good understanding of how the descendants of a single cell can interact to become the several integrated parts of a plant. Let us follow these lines to their fusion.

The trail leading from the Darwins to the demonstration of control by hormones has been followed by Mordecai Gabriel and Seymour Fogel (1955). Some of the high points follow. By 1910 P. Boysen-Jensen had demonstrated that the growth-agent effect could pass through a layer of gelatin. He and later A. Paal (1919) had cut off the light-sensitive tips of growing oat seedlings, added a drop of gelatin to the stump, and reset the tip. The growth-controlling effect could pass through the gelatin.

The next step, an elegantly simple one, was taken by Fritz W. Went (1927). He had cut off a number of apical tips of oat seedlings, placed them close together on gelatin for an hour, and then he had taken pieces of the gelatin and placed them on one side of stumps of seedlings from which the apical tips had been removed. The cells on the side directly below the gelatin elongated more rapidly than those on the uncovered side, and the seedling curved as it would had it been lighted unevenly. A plant hormone had been demonstrated.

The material produced by the growing tip was called apical auxin. As biological and chemical experiments proceeded, it was learned that several known substances could act like apical auxin. One of them—indole-3-acetic acid, IAA—is produced in growing shoots and leaves. It seems to be the major apical auxin. William Jacobs (1956) has studied just how much IAA must be applied to substitute for the action of leaves after their removal. He and his collaborators have worked with the *Coleus* plant. When leaves are growing above a given region where a piece containing xylem tubes is cut away, some of the nearby parenchymal cells line up and transform to xylem tubes.

Without the leaves, the transformation cannot occur, but a 1 percent suspension of indole acetic acid in lanolin applied to a cut stem above the region cut away replaces the effect of the leaves (Jacobs, Danielson, Hurst, Adams, 1959). Xylem tubes regenerate under the influence of IAA as well as they do with leaves present. The number of tubes depends upon the concentration of the applied IAA.

This seems like an ideal correlating system. Faster-growing apices and leaves get more vascular tissue to nourish them by producing more IAA. After much research by many investigators, it now appears that all parts of the plant differentiate and grow better because of IAA coming to them from the apex and leaves.

We also know that besides general correlative stimulation, there is specific correlative inhibition (R. Snow, 1937). As Vöchting and McCallum had established earlier, many regions are potentially apices of potentially roots. Once the two ends of a plant embryo are established, it appears that an agent suppressing shoots is acting in one direction and another agent suppressing roots is acting in the opposite direction.

Kenneth Thimann and Folke Skoog (1934) suggested that the apical hormone, in addition to being a general stimulator, was also a specific inhibitor. They applied a partially purified preparation of apical auxin to decapitated shoots and learned that the growth of the lateral buds was inhibited. When pure preparations of IAA became available, it became clear that IAA was not a specific inhibitor. Jacobs, Danielson, Hurst, and Adams (1959), using the same concentration of IAA which controlled xylem regeneration at a normal level and acted to cause normal abscission of lower leaves, found no inhibition of lateral shoots. The specific inhibition had to result from something else. There were other experiments which indicated that IAA by itself could not be the inhibitory agent. Snow (1937) knew that the agent inhibiting lateral shoots could pass up those shoots whereas auxin passes down.

P. C. Prévot (1939) demonstrated very clearly that apical auxin is not the agent suppressing the formation of buds on *Begonia* leaves. These buds as noted above can produce whole plants. As long as the leaf remains part of the plant the buds do not appear. They are apparently suppressed by something else coming from the rest of the plant. Prévot cut off everything from the base of a leaf except the fibrovascular bundles. Still no buds appeared. The suppression seemed to depend upon this transport system. George Avery (1935) had already shown that IAA moves in the opposite direction—from the leaf down the petiole. Still Prévot checked the effect of IAA on his isolated leaves. If it were really true that IAA suppressed early development of shoots, added IAA should have prevented the appearance of buds on detached leaves. Instead, the reverse happened. Many plant buds appeared on the isolated leaves.

Prévot also collected in agar what might have been coming from plant through petiole to leaf. Again there was no suppression. Not only was IAA not reaching the leaf by way of the petiole, but nothing could be collected from the cut petiole which suppressed buds. The experiment indicates either that there is no inhibitory material coming from the plant or that sufficient amounts were not collected.

It is now known that other substances besides IAA such as giberellic acid and kinetins serve as hormones, sometimes interacting with IAA and producing widespread effects. For example, the work of many investigators makes it clear that hormones, possibly blends of hormones, produced in leaves can work back on a vegetative shoot and cause it to transform to a flower. Presumably this involves turning off one set of genes and bringing another set into action.

These widespread effects, such as the switch from vegetative axis to flowering axis, like metamorphosis in animals, require coordinated genetic action throughout the organism. Wherever widespread changes are observed, hormones spreading throughout the organism are involved. In the transformation from tadpole to frog, one of the hormones at work throughout the body is thyroxin. As a larval insect metamorphoses to an adult, there are changes in proportions of ecdysone and juvenile hormone. These spectacular overall changes are under hormonal control. However, in all organisms there are cell-to-cell controls operating during differentiation. The agents of these controls are not widespreading hormones.

The work in plants has been concentrated on hormonal effects, but there is evidence that fine localized controls are also at work. Here we return to the work of Prévot. When the *Begonia* leaf is isolated from the plant and thereby removed from the effect of the general hormones, it still exhibits a kind of polarized control within itself. Epidermal cells near the main vein at the base of the leaf transform to a whole plant. As this occurs, other potential plants fail to appear. Prévot could cause new plants to appear in other parts of the leaf by surrounding those parts with an N_2 atmosphere.

Another way to do it, Prévot found, was to coat the leaf with IAA. Many buds appeared. The best way to cause many new plants to arise on one leaf was to enclose the leaf in a small chamber where it had light during the day and darkness at night.

These observations led Prévot to two important conclusions. Since K. Goebel's work (1902), the tendency had been to believe that one center of differentiation dominated its environs and blocked regions otherwise capable of the same type of differentiation by successfully competing for some substrate in short supply. Prévot observed that there was plenty of whatever was needed to make many plants. Instead of control operating by way of competition, a differentiating region acted by direct repression.

Both N_2 and auxin acted to block the control of the differentiating region. Prévot, in agreement with the Child (1929, 1941) theory of metabolic gradients, suggested that normally one small part of the leaf can become the active center. N_2, by depressing metabolism and auxin by elevating it all over the leaf, could make a number of regions sufficiently alike so that many could act as centers of differentiation. This kind of relationship has not been thoroughly studied in plants, but we shall return to it in the chapter on hydroids.

Mary and Robert Snow (1948) studied the positioning of leaves and realized that each leaf primordium must in some way suppress nearby adjacent cells from becoming leaf. C. W. Wardlaw (1950, 1965), went on to study this experimentally. He made isolating cuts on parts of growing and differentiating apices of ferns and flowering plants.

The apex of a plant consists of a central growing region or meristem. This region of cell division may consist in different species of one to many dividing cells. Some of the peripheral daughter cells left behind begin to shape up as leaf primordia. The youngest leaf primordia are closest to the ever-growing center. As the leaves differentiate and grow, they come to lie farther and farther away from the center. The youngest leaves always appear close to the apical meristem.

When Wardlaw made an isolating cut between an active apical meristem and the youngest leaf primordium, the cells on the far side of the cut, away from the apical meristem, themselves became apical meristem and a two-headed plant developed.

There has always been the tendency for the first attempt at explanation of this kind of phenomenon, wherever found in plant or animal, to be based on the notion of competition (Wardlaw and Cutter, 1955). In this case it was thought that possibly the apical meristem was preventing the leaf primordium from remaining apical meristem by drawing scarce nutrients away from it and thus in some way forcing it to become leaf. As far as we know now, differentiation never works this way; certainly not in this case. As Erwin Bünning (1952) pointed out, there is never any change in the pattern of differentiation caused by varying nutrition. In fact, one can isolate apices in a culture where all of the substrate ingredients are known and in good supply, and instead of a lot of apical meristems forming, as one would expect in this culture of plenty, only one forms. Furthermore, these isolated apices can sometimes go on and differentiate into whole plants (Allsopp, 1964). Neither in extreme plenty nor during starvation does the correlative interaction change (Bünning, 1952). If control of differentiation is not the result of competition for substrates and if IAA is not an apical inhibitor, what other explanation is there? First, there is much evidence from studies on plants as first suggested by McCallum that shoots at one end of a polarized system may directly inhibit shoots, and roots at the other end may directly inhibit potential roots in the opposite direction. The Wardlaw studies demonstrate further specific control. The apical meristem suppressed potential apical meristems without suppressing differentiation and growth of leaves. There is even some indication that older leaves may repress the growth of younger leaves (Allsopp. 1964). In summing up localized control Bünning said

... it is not simply an active growth of protoplasm that prevents further active growth of protoplasm from taking place in the vicinity. The decisive criterion is the specificity of the process. A certain type of embryonic growth will not allow the same type of growth to take place nearby, a different type of growth, however, will proceed unimpaired."

The need for a search for specific inhibitors seems clear. Ru-Chih Huang and James Bonner (rev. by Bonner, 1965) have presented evidence that the repression is at the genetic level by RNA-histones (Chapter One).

The work on plants makes it very clear that if differentiation is the result of specific repression, the repression operates in a polarized system. Knowledge about a possible polarized transport system was presented before it was known what, if anything, was being transported. Most of this information was gained during the 1920s and 1930s by Professor Elmer J. Lund and his associates. It became clear that there were polarized electrical potential differences in all growing and differentiating systems. The work was done on plants and hydroids at about the same time and will be considered together. The dominant regions were electropositive at the surface in, for example, the Douglas fir (Lund, 1929a,b), Obelia (Lund, 1925), and onion roots (Rosene, 1934; Lund, Mahan and Hanzen, 1945). This dominant region was electropositive both to more proximal regions on the surface and to internal tissue of the same region (Lund, 1925; Schrank, 1951, 1957). Schrank's work suggests the possibility of circuits, with the potential differences in one direction on the surface and in the opposite direction internally.

The potential differences between regions are generated by regional differences in oxidation and reduction. The dominant regions have been shown to be centers of greatest oxidation-reduction activity in many organisms (summarized by C. M. Child, 1941). The regions of highest positivity show the greatest capacity for reduction of methylene blue and most rapid respiration (Lund and Kenyon, 1927). The potential differences disappear or are inverted when O_2 is removed from the mitotic zone of roots (Rosene, 1934). When O_2 is replaced, there is an abrupt temporary increase in the electromotive force to above the normal level. Substances such as KCN which depress respiration also abolish dominance (Child, 1941) and electrical differences (Rosene and Lund, 1930; Lund, 1931).

The above correlations indicate the possibility that polarized differences in oxidation-reduction rate generate electrical potential differences which control growth and differentiation.

That the electrical polarity could determine the direction of morpho-genesis was also demonstrated by Lund (1921, 1923) in hydroids. By applying a field of reversed sign, he could reverse the direction of morpho-genesis. This and the work which has resulted from it are covered more thoroughly in the chapter on hydroids.

In already differentiated pieces of plants it has not been possible to change the polarity, but it is possible to block activity by applied fields of reversed sign. There was inhibition of both cell division and cell elongation in onion root tips with a drop in potential of only 40 millivolts per milli-meter (Lund, Mahan, and Hanzen, 1945). This was true when the root tips faced the negative pole. When they faced the positive pole and were oriented with rather than against the inherent polarity, there was little or no effect. This indicates that the activities of specialized regions can be blocked by opposed currents.

Although the demonstration that polarity can be reversed by imposed fields of natural strength has not been made in plants, the correlation between respiratory differences, potential differences, and dominance is excellent. In addition, D. S. Fensom (1957-1963) has demonstrated that the magnitude of the potential differences generated by regional metabolic differences is sufficient to account for polarized transport. The flow of fluid in excised stems varies with light and temperature, and changes are correlated with and preceded by changes in electrical potential.

We shall return in the chapters on worms and hydroids, where polarity can be reversed, to one of Lund's suggestions—the possibility of directed transfer between cells in contact, of substances "unknown" at the time of his writing.

There has been considerable confusion in the botanical literature concerning directed transport control. One of the reasons is that fields of greater than normal strength, whether in line with or opposed to the bioelectric field, stop protoplasmic streaming within cells. Because such strong fields stop all transport, the transport could not be recognized as polarized in an imposed electrical field. As H. G. DuBuy and R. A. Olson (1938) explained, much of the transport is by protoplasmic streaming within cells. In which direction charged substances can leave a cell would depend upon the sign at the cell surfaces. Each cell junction may be thought of as a valve which allows passage in one direction only.

Another difficulty in understanding polarized transport control arose from the assumption that indole-acetic-acid is the only controlling sub-stance transported. In some cases the control spread in one direction and the IAA in the opposite direction.

A third difficulty was the inconstancy of electrical sign. Stimulation and injury as noted above can reverse the sign.

In spite of the difficulties encountered, the general picture of control of differentiation in plants is clear. All cells contain the information enabling them to climb the ladders of differentiation all the way to tip of shoot or tip of root status. Although the cells lying behind them carry the same information in their DNA codes, they cannot use it unless isolated from cells closer to the end of the system. The effective repressive control spreads from tips toward the center. The repression may be at the DNA level and it may be accomplished by specific RNA-histones. It now becomes important to know whether these are transported from cell to cell and whether RNA-histones released by apical cells prevent like differentiation in potentially like cells.

It should be noted that although fewer investigators have worked with plants than with animals, the plants have been such favorable material that general relationships have usually been first understood by botanists.

References Cited

Allsop, A. 1964. Shoot morphogenesis. *Ann. Rev. Plant Physiol.* 15:225–254.

Avery, G. 1935. Differential distribution of phytohormone in the developing leaf of *Nicotiana* and its relation to polarized growth. *Bull. Torrey Bot. Club* 62:313–330.

Bonner, James. 1965. *The Molecular Biology of Development.* 155 pp. Oxford and New York: Oxford University Press.

Boysen-Jensen, P. 1910. Über die Leitung des photropischen Reizes in Avenakeimpflanzen. *Berichte d. Deutsche Gesellschaft* 28:118–120.

Bünning, E. 1952. Morphogenesis in plants. In *Survey Biol. Progr.* 2:105–140.

Child, C. M. 1929. Physiological dominance and physiological isolation in development and reconstitution. *Arch. Entw-mech. Org.* 117:21–66.

———. 1941. *Patterns and Problems of Development.* Chicago, Ill.: University of Chicago Press.

Darwin, C. and Darwin, F. 1880. Sensitiveness of plants to light: its transmitted effects. From *The Power of Movement in Plants.* London.

de Candolle, A. P. 1832. Physiologie vegetale T. III, p. 1474.

DuBuy, H. G. and Olson, R. A. 1938. Protoplasmic streaming and dynamics of transport through living cells. *Biodynamca 2*, 36:1–18.

Fensom, D. S. 1957. The bioelectric potentials of plants and their functional significance. I. An elecktrokinetic theory of transport. *Can. J. Bot.* 35:573–582.

———. 1958. II. The patterns of bioelectric potential and exudation rate in excised sunflower roots and stems. *Can. J. Bot.* 36:367–383.

———. 1959. III. The production of continuous potentials across membranes in plant tissue by the circulation of the hydrogen ion. *Can. J. Bot.* 37:1003–1026.

———. 1962. Changes in the rate of water absorption in excised stems of *Acer saccharum* induced by applied electromotive forces: the "flushing effect." *Can. J. Bot.* 40:405–413.

———. 1963. Some daily and seasonal changes in the electrical potential and resistance of living trees. *Can. J. Bot.* 41:831–852.

Gabriel, M. and Fogel, S. 1955. *Great Experiments in Biology.* Englewood Cliffs, N. J.: Prentice-Hall.

Goebel, K. 1902. Uber Regeneration im Pflanzenreich. *Biologisches Centralblatt* 22:385–397.

Jacobs, W. P. 1956. Internal factors controlling cell differentiation in the flowering plants. *Amer. Nat.* 90:163–169.

Jacobs, W. P., Danielson, J., Hurst, V., and Adams, P. 1959. What substance normally controls a given biological process? II. The relation of auxin to apical dominance. *Devel. Biol.* 1:534–554.

Lund, E. J. 1921. Experimental control of organic polarity by the electric current. I. Effects of electric current on regenerating internodes of *Obelia commisuralis. J. Exptl. Zool.* 34:471–493.

———. 1923. Threshold densities of the electric current for inhibition and orientation of growth in Obelia. *Proc. Soc. Exp. Biol. Med.* 21:127–128.

———. 1925. Experimental control of organic polarity by the electric current. V. The nature of the control of organic polarity by the electric current. *J. Exptl. Zool.* 41:155–190.

———. 1929a. Electrical polarity in the Douglas fir. *Publ. Puget Sound Biol. Stat.* 7:1–28.

———. 1929b. The relative dominance of growing points in the Douglas fir. *Publ. Puget Sound Biol. Stat.* 7:29–37.

———. 1931. The unequal effect of O_2 concentration on the velocity of oxidation in loci of different electrical potential, and glutathione content. *Protoplasma.* 13:236–258.

Lund, E. J. and Kenyon, W. A. 1927. Relation between continuous bioelectric currents and cell respiration. I. Electric correlation potentials in growing root tips. *J. Exptl. Zool.* 48:333–357.

Lund, E. J., Mahan, R. I. and Hanzen, A. H. 1945. Electric control of polar growth in roots of Allium cepa. *Proc. Soc. Exptl. Biol. Med.* 60:326–327.

McCallum, W. B. 1905. Regeneration in plants. *Bot. Gaz.* 40:97–120, 241–263.

Paal, A. 1919. Uber phototropische Reizleitung. *Jahr. f. wiss Bot.* 58:406–458.

Prévot, P. C. 1939. La néoformation des bourgeons chez les végétaux. *Mem. Soc. Roy. Sci. Liège*, 4 serie 3:175–342.

Regel F. 1876. Die Vermehrung der Begoniaceen aus ihren Blättern. *Jen. Zeitschr. für Naturwissenschaft* 10:447–492.

Rose, S. M. 1952. A hierarchy of self-limiting reactions as the basis of cellular differentiation and growth control. *Amer. Nat.* 86:337–354.

Rose, S. M. 1960. A feedback mechanism of growth control in tadpoles. *Ecology* 41:188–199.

Rose, S. M. and Rose, F. C. 1961. Growth-controlling exudates of tadpoles. In "Mechanisms in biological competition," ed. F. L. Milthorpe, *Symp. Soc. Exp. Biol.* 15:207–218.

———. 1965. The control of growth and reproduction in freshwater organisms by specific products. *Mitt. Internat. Verein Limnol.* 13:21–35.

Rosene, H. F. 1934. Dependence of continuous bioelectric currents upon cell oxidation. *Proc. Soc. Exp. Biol. Med.* 31:687–689.

Rosene, H. F. and Lund, E. J. 1930. Evidence from the effects of KNC for the linkage between polar growth, electric potentials, and cell oxidation. *Publ. Puget Sound Biol. Stat.* 7: 327–334.

Schrank, A. R. 1951. Electrical polarity and auxins. In *Plant Growth Substances* (Folke Skoog, ed.) pp 123–140. Madison: Univ. of Wisconsin Press.

———. 1957. Bioelectrical implications in plant tropisms. *Symp Soc. Exp. Biol.* 11:95–117.

Snow, M., and Snow, R. 1948. On the determination of leaves. *Symp. Soc. Exp. Biol.* 2:263–275.

Snow, R. 1937. On the nature of correlative inhibition. *New Phytologist* 36:283–300.

Sterling, C. 1950. Histogenesis in tobacco stem segments cultured *in vitro*. *Amer. J. Bot.* 37:464–470.

Steward, F. C. 1961. Growth induction in explanted cells and tissues. Metabolic and Morphogenetic Manifestations. In *Synthesis of Molecular and Cellular Structure* edit. D. Rudnick. pp. 252. New York: Ronald.

Steward, F. C., Mapes, M. O., and Mears, K. 1958. Growth and organized development of cultured cells. II. Organization in cultures grown from freely suspended cells. *Amer. J. Bot.* 45:705–708.

Thimann, K. V., and Skoog, F. 1934. On the inhibition of bud development and other functions of growth substance in *Vicia Faba*. *Proc. Roy. Soc.* B114:317–339.

Vöchting, H. 1885. Uber die Regeneration der Marchantieen. *Jahr. f. wiss. Bot.* 16:367–414.

Wardlaw, C. W. 1950. Experimental and analytical studies of Pteridophytes. XVII The induction of leaves and buds in *Dryopteris cristata* Druce. *Ann. Bot.* 14:435–455.

Wardlaw, C. W. 1965 The morphogenetic role of apical meristems: fundamental aspects illustrated by means of shoot apical meristem. *In Encyclopedia of Plant Physiology* (*Handbuch der Planzenphysiologie*), pp. 443–451. Berlin: Springer.

Wardlaw, C. W., and Cutter, E. G. 1955. Experimental and analytical studies of Pteridophytes. XXX. Further investigations of the formation of buds and leaves in *Dryopteris cristata* Druce. *Ann. Bot.* 19:515–526.

Went, F. W. 1927. On growth-accelerating substances in the coleoptile of *Avena sativa*. *Proc. Kon. Akad. Weten Amsterdam* 30:10–19.

CHAPTER FIVE

Protozoa–Polarized Control of Differentiation within the Cortex of Single Cells

The problem of cellular differentiation is essentially a question of how groups of cells all with the same genes can react with each other so that different parts of that genome are used in the various cells. In the Protozoa, the question is how differences can arise in a single cell with a single genome. One would hardly expect to study cellular differentiation in single-celled organisms, but the work on control of enzymatic expression in *E. coli* (Chapter one) and the works on *Paramecium* and *Stentor* in this chapter have done much to shape the modern views of differentiation.

Tracy Sonneborn (1948) has clearly demonstrated that genetically identical paramecia can differentiate. He dealt with two stocks of *Paramecium aurelia*. These two stocks differed in some of their genes, but within each stock the individuals were identical genetically. In spite of genetic identity, it was possible to recognize a series of differences within a stock by serological tests. When individuals from one subculture of stock 51 were injected into a rabbit, an antiserum was later found in the rabbit's blood. This, when used on paramecia of the same sub-culture, killed them. When the same dilution of the antiserum was used for the same time on other subcultures, some had individuals which were not killed. Still other subcultures were like the first in that the individuals were killed by the antiserum. All cultures composed of paramecia sensitive to this antiserum were denoted A. Paramecia insensitive to the A-antiserum were, however, killed by antibodies prepared against members of their own subculture. In this way a number of serotypes, A, B, C, D, E, G, were recognized in stock 51. All of these were genetically identical but each had different antigens against which rabbits could produce different antibodies.

Another stock, number 29, of *Paramecium aurelia* had a different set of serotypes, A', B', C', D', F, H. Here again one genome could be expressed in a variety of ways.

E and G were found only in stock 51 and F and H only in stock 29. The primes indicate that in stock 29 there were some antigens similar to those in stock 51. Anti-A, for example, could affect A' animals, but the dosage had to be greater..

When there was conjugation between individuals of the two stocks resulting in the nucleus of one being transferred to the cytoplasm of the other, the old antigen disappeared. A new one, always one in the series determined by the nuclear donor, was established. This showed that the serotype was determined by nuclear genes (Sonneborn, 1950).

It was possible to rid a paramecium of its serotype by a short treatment with antibody against its antigen or by treatment with antibody against the similar antigen of the other stock. For example, D' individuals treated with anti-D serum for a short time and washed free of it before they could be killed, lost their D' antigen. Their daughter cells showed no D' or any other antigen for some time. These daughter cells could then become the same or another serotype, provided it was one of the series for which the nucleus carried the code.

Which serotype was to be expressed depended upon the environment. In one experiment after D' individuals were treated with anti-D serum, the survivors were divided into two groups. One group was kept at 32°C and excess food was present. The others were cultured at 20°C with only enough food to allow one fission per day. Ninety-four percent of the much food-high temperature group transformed to B'. Ninety-six percent of those kept at the lower temperature with restricted food became H. Individuals of any serotype can change to any other in one or two steps. For example, Ds may first transform to Bs then to Fs.

The treatment by antiserum is not necessary in effecting a transformation, but it greatly speeds the process. If paramecia of serotypes A, B, C, D, and E were kept at 32°C, A remained A and the others became A. At 12°C all eventually became B. Once transformed, paramecia retain their new type as long as conditions remain the same.

The antigens seem to be at or near the cell surface, because the antisera quickly immobilize the surface cilia.

Here in this single-celled animal is cellular differentiation. Cells with the same genome can be called upon by the environment to use part of that genome. Furthermore, a changed environment may cause another part of the genome to be expressed. The use of one gene in the production of an antigen seems to preclude the use of another. All of these features of differentiation in paramecia including the presence of different surface antigens are also features of cellular differentiation in multicellular forms.

If differentiation were simply a case of different cells using different portions of their DNA, one would not expect to find differentiation within a single cell. Work on Protozoa, especially on *Stentor*, however, has taught us much about differentiation and much of this knowledge may be generally applicable.

Stentor coeruleus is a large ciliate about one-half millimeter long. Its complexity is apparent in Fig. 5-1. It is also an excellent regenerator. If

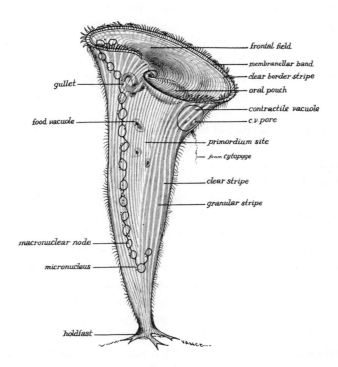

FIG. 5-1. THE ANATOMY OF STENTOR [*From Vance Tartar*, The Biology of Stentor. *Pergamon Press, 1961*].

either an anterior or a posterior portion or both are removed, the missing parts are replaced, usually in a fraction of a day. Most of our knowledge about the intracellular interactions during regeneration comes from the surgical works of Paul B. Weisz and Vance Tartar.

There are micronuclei and a long beaded macronucleus in *Stentor*. As in other ciliates, the macronucleus controls the cells' non-reproductive activities. At least one macronuclear bead of the total of 15 or more must be in a fragment before regeneration can occur. Although the macronucleus is a multiple structure, there is no evidence that parts of it control special regions. One bead, and apparently any bead, with associated endoplasm and cortex is sufficient to produce a complete individual including a fully reconstituted many-beaded macronucleus.

Surrounding the nucleus is a viscous endoplasm and the more rigid cortex. Regeneration requires part of the cortex, but most of the endoplasm can be removed.

The most anterior part of a decapitated stentor begins to transform to the missing head in a few hours. The most prominent structure and the first to appear is the relatively thick rim of the funnel with its membranelles instead of thinner cilia. This is known as the oral primordium, and the first sign of its appearance under the dissecting microscope is a break in the striping pattern. This change always takes place in a few thin stripes adjacent to the widest stripe. The oral primordium appears a short way back from the anterior end and migrates anteriorly as it curves and grows. At the same time, the entire anterior region changes shape and is remodeled until all of the missing part of the pattern has been reconstituted. If part of the old anterior structure remains, it may fuse with the new to complete a whole pattern (Fig. 5-2).

Early in the analysis of regeneration of Stentor (rev. by Tartar, 1961), it became clear that there is a special part of the cell in which the new funnel primordium appears. If this region is removed, some other region can take over and become funnel. Also the farther back the cut removing the anterior parts is made, the farther back this primordium will form. In fact, it can form at any level of the cell but that level must be close to the new anterior end. Weisz conceived of a hierarchy of control with the head in some way preventing other potential head-forming regions from achieving head status. Because the head-forming area is close to a special kinety, he thought of the anterior part of this as taking precedence over more posterior parts. When an anterior part was removed, a more posterior part of this longitudinal locus could take over. If all of the longitudinal locus of head formation was removed, another meridional locus could become head-

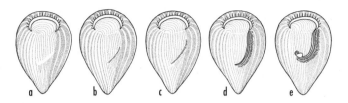

FIG. 5-2. REGENERATION IN STENTOR AFTER REMOVAL OF PART OF THE ANTERIOR
STRUCTURE [*adapted from Vance Tartar (1961)*].

a: A part of the anterior structure had been cut away a few hours before.
Already in the narrow stripe region adjacent to the wide stripes there is disruption of stripes.

b and *c*: A new oral primordium forms and grows.

d and *e*: The new oral promordium fuses with the remnant of the old structure
and forms all which is not already present. Had none of the old remained, the
new would have formed all.

forming. This usually required considerably more time to get started
(Weisz, 1954, Tartar, 1961) than did the special locus where the widest
stripe lay alongside the field of thinnest stripes.

Tartar (1956a, b) learned what is special about the head-forming locus.
By grafting stentors together in various ways and by cutting around a piece
of cortex, rotating it in position so that stripes of different width abutted, he
learned that wherever there was a locus of sudden change in width of stripes,
there a new head primordium could arise (Fig. 5-3). These new centers of
organization could be created at will. This clearly indicated that there is no
special head-forming locus but that any cortical region could become the
center of organization of a head. In the case of two loci of great stripe
contrast, two centers of organization developed from which differentiation
spread to result in almost complete doublets.

The initiation of such a center of organization appears to have no
special preformed substance and need have no remnant of old head structure.
It appears wherever the experimenter chooses to make a region of great
contrast in stripe width. If two areas of great contrast are produced near
each other, the one with greater contrast is successful and inhibits the one
of lesser contrast (G. Uhlig, 1960).

The above works tell us that new pattern is established during regeneration and that it is not by completion of the old pattern. Rather, any part of
the surface—provided contrast has been provided and that a macronucleus

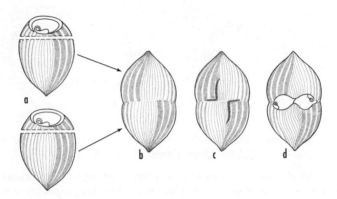

FIG. 5-3.   A demonstration that new oral primordia appear at loci of abrupt change in width of stripes [*adapted from Tartar (1961)*].

a: Two anterior ends are removed.

b: The remaining bodies are fused anterior to anterior.

c: Two L-shaped loci of great contrast in stripe width are produced, and at each a new oral primordium forms.

d: The two primordia become two new anterior ends.

is present—can become a head *de novo*. The old is not a template for the new.

What is it about a head that prevents other potential head regions from asserting themselves? If a head is removed, a new one forms from the anterior part of the decapitated stentor. This does not happen when a head is removed and it or another is grafted back in the anterior site (Tartar, 1958). Also, a regenerating head regresses if a stentor with a head is grafted anterior to it (Tartar, 1964).

It is quite clear now that head can inhibit potential head, but it is not quite that simple. The head to inhibit must lie anterior to the region being inhibited. If a head is removed, regeneration of a new one is expected (Fig. 5-4[a]) As noted above, an anterior graft with a head can prevent this in the posterior partner (Fig. 5-4[b]) Although head can inhibit potential head, the anterior partner may have its head removed and still prevent the posterior partner from forming a head (Fig. 5-4[b′]). As we now know from the work on stentors and also from the work on worms and coelenterates (Chapters six and seven) any more anterior level can prevent a more posterior level from achieving head status.

Another generalization is that control spreads primarily in one direction, in this case from anterior to posterior. If one starts with 4 decapitated individuals, and two are put together with posterior of one to the anterior end of the other, and if the other pair are fused anterior to anterior, the results are very different. In the first pair, the anterior member regenerates a head and the posterior member does not. In the identical pair, but fused anterior to anterior, both produce heads (Fig. 5-4[c]). This tells us that the posterior member is not blocked in head formation by having its wound sealed. When that wound is sealed by another but with opposite polarity, no control passes from one to the other. Control spreads only from anterior to posterior. When the two are together anterior to anterior, a divide lies between and the streams down the two sides of the mountain are unmixed.

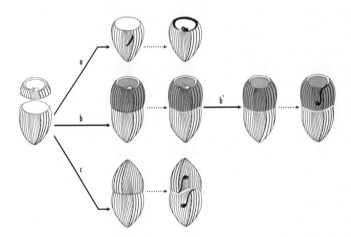

FIG. 5-4. POLARIZED CONTROL OF ANTERIOR OVER MORE POSTERIOR LEVELS [*adapted from Tartar 1961, 1964*)].

a: New anterior structures form after removal of the original anterior structures.

b: If after removal of anterior structures, as in *a*, another anterior region is grafted anterior to the piece in question, no new anterior structures will appear in the posterior partner.

b': Even after removal of the anterior structures from the anterior partner, new anterior structures do not form in the posterior partner. The anterior member, even without its most anterior structure prevents the regeneration of the anterior structures in the posterior partner.

c: When two bodies minus their anterior structures are fused anterior to anterior, both regenerate anterior structures. Repressive control is polarized, and is only effective when one lies behind another.

The control system is more complicated than the simple analogy of a watershed. One way Tartar showed this was by starting with an old observation by S. O. Prowazek (1904). He cut a dividing *Stentor* through the primordium of the new head for the posterior member. Division like regeneration entails the formation of a new head before separation of the daughter cells. Somehow the rule that a second head cannot form behind the first is suspended for a few hours in preparation for transverse fission.

When Prowazek made his cut through the head primordium, each of the two isolates contained half a regenerating head. In the anterior member already with a head, the half primordium regressed (Fig. 5-5[a]). In the posterior isolate without a head the half primordium became a whole head (Fig. 5-5[b]). One would guess that the anterior isolate had too much head for its size and that the more posterior one regressed. Tartar continued the work and his results show that the presence of the old head does prevent the development of the new one. When the same operation was performed with a cut through the new head primordium of a divider but in addition with the original head cut away the half primordium could survive and form a whole head in both isolates (Fig. 5-5[a'] and [b]). The next operation showed something else about the control. This time Tartar took his anterior isolates with a half primordium, cut through the cortex below the head, and rotated it 180° so that the ventral part of the head fused to the dorsal part of the body. Here also the control of head over primordium behind it failed. As Tartar has emphasized, "not the materials of the organelles but their proper pattern and relationship to the whole is essential to their inhibitory effect." Certainly this rules out any specific inhibitors which diffuse freely.

The nature of the inhibitory control has not been worked out in *Stentor*, but these experiments clearly indicate that it operates over a polarized system. The major direction of control is anterior to posterior, but there is transverse asymmetry as indicated by the failure of control when a head is rotated 180° around the major axis.

Although control operates from anterior to posterior, a rearrangement in the posterior region can affect the lines of control anterior to it. For example, extra tails grafted to the side of the body may cause extra mouths to form where lines drawn parallel to the main axes of the tails would cross the posterior part of the primordium-forming area (Uhlig, 1960; Tartar, 1961).

Control can also pass from one individual to another when they are grafted side by side. In these parabiosed animals, an appreciably larger member of the pair usually exercises inhibitory control. There are also cases in which a larger animal can induce a smaller one joined to it to divide

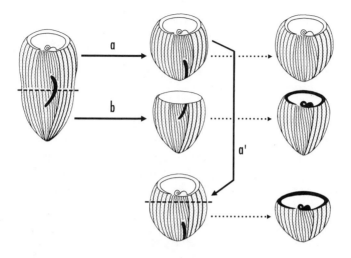

FIG. 5-5. A DEMONSTRATION THAT OLD AND NEW ANTERIOR STRUCTURES CANNOT
PERSIST IN A DIVIDER OF REDUCED SIZE [*adapted from Tartar (1961)*].

A divider with old anterior structures and a new oral primordium which would
have become the anterior part of the posterior daughter individual was cut
through the oral primordium.

a: The anterior half contained the old oral structures plus half of a new primordium: the primordium regressed.

b: The posterior half contained half a primordium and no old anterior region:
the half primordium became a whole anterior apparatus.

a′: The anterior half primordium persisted and formed a whole anterior
apparatus when the old anterior structures were removed.

prematurely (Weisz, 1956; Tartar, 1966). This type of control requiring a
larger individual to affect a smaller is less efficient than antero-posterior
control where even a smaller anterior graft can control the host behind it.
The more efficient anteroposterior control could be expected to prevail
under natural conditions.

There are other things about interaction in *Stentor* which are mentioned
not because we can explain them now but because they may be clues to
where valuable information may be sought. If a graft is placed across the
major axis of a host so that the stripes of graft and host are at right angles to
each other, the stripes will slowly over a day or so swing around and become
parallel to those of the host. Not only are the individual parts of the cortex
polarized, but they can turn to line up with the host (Tartar, 1961). One is

struck by the similarity of behavior to magnetized needles affected by the earth's field (See discussion by Weisz, 1956).

There are also regional differences along the length of the animal. If a cut is made across the middle, the stripes and apparently the kineties rejoin. If, however, the posterior part of one stentor is joined to the anterior of another in a tandem graft, the stripes do not join but groups of them over a period of several days slide slowly between groups of stripes of the other. Those of the partners come to lie side by side as those of the anterior partner move posteriorly, and those of the posterior partner in the opposite direction The result is an unusually wide stentor with too many stripes around its circumference. Eventually some of the stripes disappear and there is regulation toward normal dimensions (Tartar, 1964).

Returning again to parts of the cortex acting as centers of organization, this has also been observed in *Paramecia*. Sonneborn (1963) found a Paramecium which had received during conjugation a piece of its partner's paroral cortex. When this individual prepared to divide there were two centers of new mouth formation. Likewise when the daughter cell divided two more primordia formed. The result was a clone of double-mouthed *Paramecia*.

This is in a sense inheritance of an acquired mouth, but it is a peculiar kind of inheritance. It turns out that it does not make any difference whether affected individuals received nuclei or cytoplasm from non-affected individuals. Their daughter cells are partially doubled if the cortex is partly doubled. The inheritance of the partially doubled structure depends only upon having an extra oral apparatus and the associated abnormal cortical pattern.

When the new oral primordia appear, very much as they do in *Stentor*, they are close to the old mouths. However, as Sonneborn has pointed out, "between the old and the newly formed parts there is no evidence of a simple direct, template-like, causal relationship."

We already know from the works of Tartar and of Uhlig on *Stentor* that new oral structures develop where the cortical pattern changes abruptly and no old mouth need be in the organism for this to occur. So it really is not a case of inheritance of an extra mouth but rather inheritance of a region from which pattern spreads and a mouth is part of that pattern. We may still think of the chromosomal DNA carrying the code for the ultimate proteins and other substances which fit in particular ways to make micro- and macro-structure. The cortex by its changing structure seems able to provide one or more spots where assembly begins.

Differentiation in multicellular organisms is at least partially the result of different portions of the DNA operating in different cells. Differentiation in single-celled organisms seems to depend upon the ultimate products of different genes taking different positions in the same cell. Is this kind of differentiation also found in cells of multicellular organisms? We shall see that it is in fertilized eggs (Chapter Eight). There is also a similarity in the polarized differentiation system of ciliates and the system working across cell boundaries in hydroids (Chapter Seven).

References Cited

Prowazek, S. 1904. Beitrag zur Kenntnis der Regeneration und Biologie der Protozoen. *Arch. Protistenk.* 3:44–59.

Sonneborn, T. M. 1948. The determination of hereditary antigenic differences in genically identical Paramecium cells. *Proc. Nat. Acad. Sci.* Washington 34:413–418.

———. 1950. Cellular transformations, pp. 145–164. In *The Harvey Lectures*, Series 44. Springfield, Illinois: Charles C. Thomas.

———. 1963. Does preformed cell structure play an essential role in cell heredity? In *The Nature of Biological Diversity*, ed. J. M. Allen, pp. 165–221. New York: McGraw-Hill.

Tartar, V. 1956a. Grafting experiments concerning primordium formation in *Stentor coeruleus*. *J. Exptl. Zool.* 131:75–122.

———. 1956b. Further experiments correlating primordium sites with cytoplasmic pattern in *Stentor coeruleus*. *J. Exptl. Zool.* 132:269–298.

———. 1958. Specific inhibition of the oral primordium by formed oral structures in *Stentor coeruleus*. *J. Exptl. Zool.* 139:479–505.

———. 1961. Activation and inhibition of the oral primordium. Chapt. 8. In *The Biology of Stentor*, pp. 135–158. New York, Oxford: Pergamon Press.

———. 1961. The Biology of Stentor. pp. 413. New York: Pergamon Press.

———. 1964. Morphogenesis in homopolar tandem grafted *Stentor coeruleus*. *J. Exptl. Zool.* 156:243–252.

———. 1966. Stentors in dilemmas. *Zeit. f. Allg. Mikrobiologie* 6:125–134.

Uhlig, G. 1960. Entwicklungsphysiologische Untersuchungen zur Morphogenese von *Stentor coeruleus*. Ehrbg. *Arch. Protistenk.* 105:1–109.

Weisz, P. B. 1954. Morphogenesis in protozoa. *Quart. Rev. Biol.* 29:207–229.

Weisz, P. B. 1956. Experiments on the initiation of division in *Stentor coeruleus*. *J. Exptl. Zool.* 131:137–162.

CHAPTER SIX

Regeneration in Worms—Neural Control of Polarized Gradients of Differentiation

So far, we have learned that in salamanders, plants, and ciliates the operation of a developmental sequence can begin in a number of regions after actual isolation or after virtual isolation as the result of pattern discontinuity. It has been suggested that the developmental sequence is a translation of a genetic hierarchy. It seems that a functional member of the hierarchy in a more distal position can specifically inhibit the like activity in regions proximal to it. The control is always polarized. Experiments on worms and hydroids have taught us most about this polarized system and the kind of information passing over it.

One basic general relationship was noted and thoroughly studied by Charles Manning Child (rev. 1941). In many organisms, the rate of regeneration of a part is greatest at some center and grades off from that center. Often the quality of regeneration also grades off from a center, as is the case of the lens noted in Chapter Two. This region, which achieved a high level of differentiation, also dominated its neighbors and prevented them from achieving the same level of differentiation. The basic correlation was that these differentiating, dominating areas had the highest general metabolic activity in their regions.

H. V. Brøndsted (1954, 1955) demonstrated this relationship in head blastemas of planaria. If he allowed a blastema to form and then cut it into medial and lateral pieces, both regions went on to form brain. The medial piece did this more rapidly. When medial and lateral regions remained together, a brain developed only in the medial region. This is an example of the general rule that a more rapidly developing region can suppress like differentiation in its neighborhood.

A gradient in regenerative ability in planaria was demonstrated by Child and Watanabe (1935). When planaria were decapitated by a cut passing just below the head, a perfect new head almost always regenerated on the anterior end of the body. As the worms were cut farther and farther to the posterior, fewer regenerated and the quality of the heads which did regenerate graded off with increasing distance from the head region. In general, as the quality and quantity of regeneration decreased, it was the anterior central part of the head which was lacking in the regenerate. Eyes tended to be closer together, or fused, or were missing. Finally at a level just behind the pharynx there was no head regeneration (Fig. 6-1)).

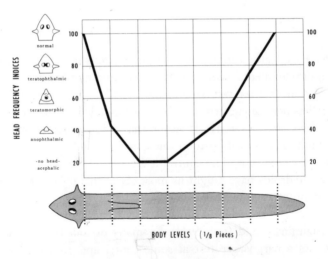

FIG. 6-1. QUALITY OF REGENERATION DIFFERS WITH BODY LEVEL IN PLANARIA
[*adapted from C. M. Child (1941)*].

Along the abscissa are different levels of the body. The quality of regeneration in arbitrary numbers and the number assigned to each type are given along the ordinate. When 1/8 pieces are isolated, the quality of regeneration falls off from perfect head regeneration just behind the old head to no head regeneration just behind the pharynx. Beyond the postpharyngeal low the ability then increases.

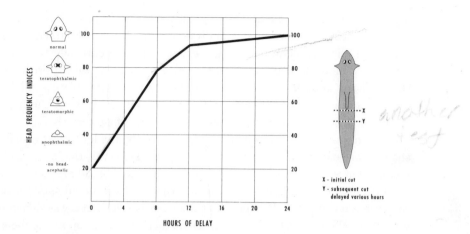

FIG. 6-2. REGENERATION OF SHORT PIECES OF PLANARIA ENHANCED BY DELAY OF SECOND ISOLATING CUT [*adapted from C. M. Child (1941)*].

No heads regenerate when isolating cuts are made at X and Y simultaneously. If the Y cut is delayed 4 hours, about half of the regenerates are teratomorphic and half are anophthalmic. Most heads regenerated are normal if the Y cut is delayed 12 hours. If delayed 24 hours, practically all are normal.

Child and Watanabe, using an arbitrary scale, expressed this decline in quantity and quality in what they called a head frequency curve (Fig. 6-2). Regeneration went from 100 just behind the head to 20 below the pharynx. They obtained this curve by cutting pieces from different levels of the body. One might think that each level was endowed with just so many cells capable of becoming head. The gradient in ability to regenerate a head would then be a gradient in number of head-forming cells. Such a suggestion has been made to explain decrease in head frequency in both worms and hydroids. It is considered in greater detail in Chapter seven. That it is not the correct interpretation was indicated by further work by Child and Watanabe on planaria.

They had learned that even the very poor head-forming level posterior to the pharynx could become just about as good as the best if the anterior cut of isolation was made 12 hours before the posterior cut (Fig. 6-2). Or it could be done the other way around. Quite good regeneration of heads occurred when the posterior end of the isolate was given a head start by cutting away the tail 4 days before the anterior cut was made.

One might still think that special regeneration cells had migrated into the piece after the first cut and were already there in the piece to be isolated before the sceond cut was made. With this in mind, John Martinek (1968). compared the regeneration of two groups of differently shaped post-pharyngeal pieces of equal volume. One group was composed of broad and short pieces, and the other contained long and narrow pieces. All had anterior surfaces which came from the same body level. Since the hypothetical regeneration cells are supposed to grade off posteriorly and since the long narrow pieces extended farther posteriorly, the long narrow pieces should have contained fewer regeneration cells. However, the long narrow pieces often regenerated heads, whereas the short and broad pieces failed to form heads. Furthermore, long narrow strips cut from broad short pieces sometimes regenerated. Having a longer antero-posterior axis was somehow important. We shall return to this after further consideration of the theory of totipotent regeneration cells.

In an earlier time when differentiation was thought to be a process dependent upon the segregation of organ-determining substances, regeneration was a problem. Numerous cases were known in which one region of an organism could produce other parts. Obviously, if as the result of segregation, a region did not contain the information for production of other parts, regeneration could not occur. Segregation theory was temporarily saved by the invention of a special line of cells with total information. These were conceived to be a line of totipotent cells like the germ-cell line but scattered throughout the body wherever regeneration could occur. Long after the segregation theory was discarded, the regeneration cells remained. In the planarians they were known as neoblasts. Some of the cells which are called neoblasts are cells preparing to divide. They have large nucleoli and much cytoplasmic RNA. There is no doubt that they take part in regeneration, but there is serious question that they are a special line of totipotent cells. Instead, it seems that these actively dividing cells usually arise from differentiated cells at regeneration sites. (Pederson, 1961; Flickinger. 1964; Woodruff and Burnett, 1965). In addition, other prominent cells which have been called neoblasts are seen to be well differentiated gland cells in electronmicrographs. (E. D. Hay, 1968). However, there are many people who believe in totipotent regeneration cells. A recent paper on the subject is that by S. Le Moigne (1966).

This does not mean that cells cannot migrate to a regeneration site from some distance and take part in regeneration. This was the case in the well executed experiments of Francoise Dubois (1949), and F. Stephan-Dubois (1965). After planaria were X-rayed with dosages great enough to prevent regeneration, they would regenerate if given grafts of unirradiated tissue.

Donors and hosts were chosen to differ in color and shade. The graft could be inserted far from the amputation surface, but the regenerate as it appeared on that surface had the shade and color of the graft. This indicated that cells can migrate to regenerating surfaces and take part in regeneration. Although the wandering cells were thought to be neoblasts, this experiment does not tell us about the past history of the cells—whether they came from a line of totipotent cells or whether they had dedifferentiated. The cells which move from unirradiated to irradiated areas are generally thought to be neoblasts. However, Teiichi Betchaku (1967) concludes from *in vitro* studies that neoblasts are poor migrators.

There seems to be no clear-cut evidence for a line of undifferentiated totipotent cells. Instead, we now believe that all cells contain total information and can under the proper conditions be shown to be totipotent. What the proper conditions are is what is being learned now. Child and Watanabe were getting close to the modern interpretation when they wrote in 1935:

"In the intact planarian the cells of all postcephalic levels are capable of giving rise to a head (in most species) but are evidently prevented from doing so by their physiological relations with other regions of the body anterior and posterior to them. When the parts anterior to any level are removed by section, the physiological factors preventing head development which originate in those more anterior parts are eliminated. Apparently these are the most effective factors in preventing head development at postcephalic levels in the intact animal, since their removal usually results in head development unless the piece is short."

So far we have considered only regeneration at anterior and posterior surfaces. When a part of the structure is missing distally, cells near the surface can transform to the missing part. In planaria, central parts are also reconstituted when lacking. Several investigators have demonstrated that if two adjacent levels are cut apart and simply grafted back together, there is no growth of new tissue between them. Only if something is missing between two pieces grafted together does new tissue form between them (rev. by Wataru Teshirogi. 1963). This means that what we are trying to understand is a kind of control that extends throughout an organism such that if anything is missing, a part of another region transforms to the missing part. The transformation to the missing part seems to result from derepression with subsequent advance up the ladder of differentiation to the highest unoccupied rung.

Regionally specific repressors have been demonstrated in three phyla of worms. Theodore Lender (1956, 1960, 1965), working in the laboratory of

Etienne Wolff (1962), has demonstrated that brain regeneration in planaria is decreased or prevented by extracts of heads but not by extracts of tails. The effect appears to be quite specific, because a brain grafted close to an anterior surface of amputation can suppress or decrease brain regeneration without suppressing the regeneration of the rest of the head. Quite the reverse is true. Once a brain region has formed autonomously in an anterior region or when a brain has been grafted to a non-anterior region (Felix Santos. 1929, 1931 and James Miller, 1938), more posterior levels may form behind it. Non-brain anterior regions may also act as centers of reorganization (Y. K. Okada and H. Sugino, 1937).

Repression of like structures and induction of non-like may be parts of the same process. Lender (1956) cut out a window from the heads of planaria, thereby removing the brain. Such brainless worms do not regenerate eyes until the brain has regenerated. "No brain, no eyes" is the rule. When windowed planaria were cultured in extracts of heads, brain regeneration could be delayed or prevented but eyes regenerated. The usual explanation is that the extracts contained inhibitors of brain and inductors of eyes. Another possibility is that brain and eyes result from the operation of different parts of the same genetic hierarchy. Then repressors of brain DNA would be allowing non-repressed DNA leading to eye formation to operate. This, however, requires the assumption that brain DNA is more easily repressed than is eye DNA by head extracts. This is an unresolved problem and one which could be tackled experimentally. The question is - if there are two members of a hierarchy, A and B, and A is repressed, is B then on automatically? If so, a repressor of A would in a sense be an inductor of B.

It might be thought from the evidence presented so far from studies on planaria that only brain is specifically repressible. However, eyes seem to be self-limiting (S. Pentz and F. Seilern-Aspang, 1961) and extracts of pharyngeal region or extracts of the region around the pharynx after removal of the pharynx can inhibit the regeneration of a pharynx (Wolff, Lender and C. Ziller-Sengel, 1964, 1967). The fact that the pharyngeal region minus pharynx can inhibit regeneration of pharynx leads us on to another relationship which has been more easily and more thoroughly studied in nemertean and annelid worms.

A worm which has proven very useful in the study of interacting regions is *Lineus vegetus*. This is a marine nemertean quite different from the planarians in form and organ structure, and it has a much more complicated body plan, but the rules of regeneration are fundamentally the same in the regenerating members of both phyla. This species of *Lineus* can regenerate heads from anterior surfaces of isolates from all levels of the body. The ability to form tails from posterior surfaces also extends throughout the body.

Marie Tucker (1959) ground up heads, centrifuged the brei and found that the supernatant fluid contained something which would inhibit head regeneration in isolated portions cultured in the fluid. While head regeneration was delayed for nine to ten days in the isolates, their other ends reshaped into tails on time. Conversely, tail supernatants delayed tail regeneration for many days without affecting head regeneration on the same pieces. It appeared to be like preventing like, head stopping potential head and tail blocking transformation to tail. Like affecting potentially like is true, but there is more to it than that. It was noted above that Catherine Ziller-Sengel had learned from planarians that an extract of the pharyngeal region even without pharynx could block the regeneration of pharynx. Here in *Lineus*, more of the nature of this kind of control between regions has been demonstrated.

Tucker prepared extracts from mid-portions of worms and tested their effects on isolates from more anterior and from more posterior regions. Posterior isolates taken from a part of the worm behind the level used for preparing supernatants did not regenerate heads in the presence of the mid-body supernatants. The same material used on anterior isolates did not delay their regeneration of heads.

This was a very clear-cut result. Mid-body supernatant divided into two parts inhibited heads on all pieces taken from behind, but it did not affect head regeneration on any anterior pieces. Pieces from the same anterior levels were blocked, however, by supernatants from heads.

The rule is that heads do not form on isolates that receive supernatants from more anterior levels. It is not simply a case of head inhibiting potential head.

Tail control works in the opposite direction and is just as regionally specific. Supernatants from tails block tail regeneration in isolates from all levels. A mid-body supernatant does not block tail regeneration in more posterior levels, but more anterior pieces fail to regenerate tails while regenerating heads on time. Here the rule is that tails do not form on isolates that receive supernatants from more posterior levels. These relationships are shown in Fig. 6-3.

We learn from the work of Marie Tucker that various levels of *Lineus vegetus* contain substances capable of preventing head formation at more posterior levels and tail formation at more anterior levels. There must be two series of substances. one preventing head development and the other preventing tail development. It appears that this is not a case of two

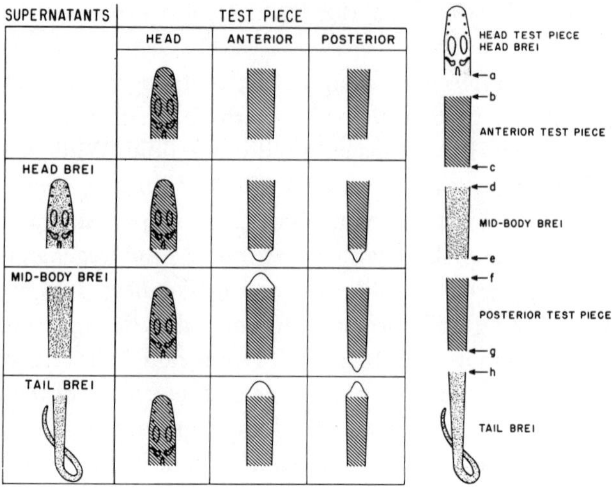

FIG. 6-3. THE EFFECTS OF AQUEOUS EXTRACT FROM DIFFERENT BODY LEVELS OF LINEUS ON THE REGENERATION OF ISOLATES FROM THE VARIOUS LEVELS [*from* Fig. 1, *Marie Tucker*, J. Morph., 105: *587*].

Extracts of head, midbody, and tail were used in culture media of heads, body pieces anterior to middle, and body pieces posterior to middle. The rule which emerges is that an extract of a head or a region between head and the test pieces blocks head regeneration while allowing tail regeneration to proceed. Working in the other direction, extracts of tails or a region between tail and the test piece stop tail regeneration without affecting head regeneration. Note the opposite effect of midbody extracts on regions more anterior and more posterior.

quantitative gradients. Tucker makes the point that the position from which the materials come, not their concentration, determines the regionality of the controlling agents. There must then be two series of qualitatively different controlling agents. However, there is a linear order such that the substance from one level can prevent all more posterior levels from achieving head status but cannot prevent a more anterior level from transforming to head. If the control is at the DNA level of organization, a geometry is suggested such that repression of one level would prevent all transcription to one side of it along the chromosome but not on the other side. Polarized transcription has been demonstrated (Benzer, 1962; Crick, 1962; rev. by Margolin, 1967).

The finely balanced regional controls seem to operate throughout the lifetime of nemertean worms except during the period when worms are fragmenting or are about to fragment prior to vegetative reproduction. Then other substances which are generally inhibitory collect in the worms primarily in posterior parts (Tucker, 1959; Reutter, 1967). These prevent all regeneration indiscriminately until they are removed,

An annelid worm has been useful for determination of the degree of regional specificity to be found in inhibitory extracts. The worm chosen was *Clymenella torquata*, with 22 segments, each at least slightly different from its immediate neighbors. Stephen D. Smith (1963, 1964), after learning that head and tail extracts inhibit regeneration specifically (Fig. 6-4), then asked whether the extract from a single segment would have the same or different effects on the regeneration of the segment just ahead of and just behind the donor of the extract. *Clymenella*, like many annelid

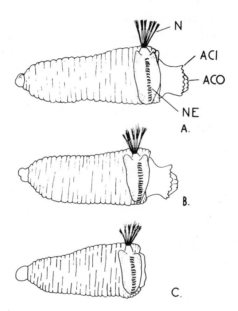

FIG. 6-4. REGIONAL SPECIFICITY OF INHIBITORS OF TAIL REGENERATION IN CLYMENELLA (*from Stephen Smith*, Biol. Bull. *Woods Hole, 125. 1963*).

a: An isolated 17th segment cultured in seawater. An anal cone, ACO, and an anal collar had regenerated.
b: An isolated 17th segment cultured in seawater plus extract of 16th segments.
c: An extract of 16th segments completely prevented all tail regeneration in 15th segments.
Whereas extracts of the segment immediately anterior to the pieces had no effect in limiting tail regeneration, extracts of the next segment closer to the tail suppressed all tail regeneration.

worms, regenerates heads well only from a few of the most anterior segments. Tail regeneration—because it occurs over a greater length of the worm, up to the middle of the body—was used in the studies on regional specificity. Smith prepared aqueous extracts from segment 16. These were added to the culture medium of isolated fifteenth and seventeenth segments. Of 28 fifteenth segments 22 healed sufficiently so that regeneration might be expected. However, 18 completely failed to show any sign of tail regeneration, and the remaining 4 made an abortive attempt. The material of segment 16 had not allowed any segment 15 to form anything resembling normal tail structures. In contrast, all seventeenth segments which remained intact did regenerate good posterior structures in the presence of sixteenth segment material.

The regional specificity was shown to be even finer. Extracts of the dorsal half of the anal collar of the last segment inhibited the regeneration of either the dorsal portion of the collar or the whole collar. Ventral extracts inhibited either the whole collar or just the ventral portion. Dorsal never inhibited ventral only, and ventral never inhibited dorsal only (Fig. 6-5). This tells us that controlling agents vary in the dorso-ventral direction as well as in the antero-posterior direction and that they are different even in parts of a part of one segment.

Again, as in the case of *Lineus*, one must suspect qualitatively different inhibitors rather than a quantitative distribution of one inhibitor. It would be too much to expect that if concentration determined specificity, the experimenter could, the first time he performed an experiment, arbitrarily choose the proper concentration of sixteenth segments which would block tail regeneration in all segments immediately ahead and not block it in any of those immediately behind.

It is concluded that each segment and even parts of segments produce different inhibitors. The next problem is how these could be distributed in such an orderly way that the many differences of an organism could arise in the same pattern in individual after individual, It has been noted in the previous chapters that the control of differentiation is always polarized. Something is known about the nature of this polarity and how it can be changed in worms and hydroids.

One pertinent point noted earlier is that there is a tendency during regeneration for the transforming region to form the most distal structures not present distal to it. This appears very clearly in the annelid, *Sabella pavonia*, from the work of N. J. Berrill and D. Mees (1936). This is a marine annelid of many segments with differentiation along the antero-posterior axis into regions called head, thorax and abdomen (Fig. 6-6).

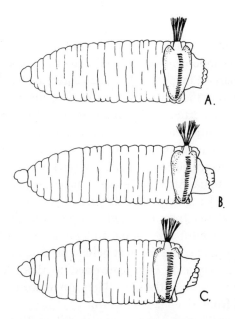

FIG. 6-5. INTRASEGMENT SPECIFICITY OF REPRESSIVE CONTROL IN CLYMENELLA
[*from Stephen Smith*, Biol. Bull., *Woods Hole, 125, 1963*].

a: An abdominal segment cultured in seawater. The posterior regenerate was complete both dorsally and ventrally.

b: An abdominal segment cultured in seawater plus an extract of the dorsal part of the last segment. The dorsal part of the regenerate is deficient.

c: An abdominal segment cultured in seawater plus an extract of the ventral part of the last segment. The ventral part of the regenerate is deficient.

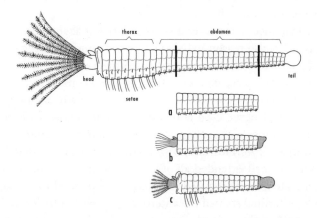

FIG. 6-6. REGENERATION IN SABELLA [*adapted from N. J. Berrill and D. Mees*, Jour. Exptl. Zool. 73: 67, 1936. *Description in the text*].

When Berrill and Mees removed the head, all of the thorax and the anterior part of the abdomen, the remaining part of the abdomen regenerated the missing parts. First, a small blastema arose on the anterior surface. One might expect that the remaining part of the abdomen would first have produced the missing portion of the abdomen, then the thorax, and finally the head, but regeneration never proceeds in that direction. It is not a case of new pattern forming as the continuation of old pattern. Instead, the small blastema transformed to a head. In *Sabella pavonia* that is all which forms from the blastema. After the head had taken shape, the more anterior remaining abdominal segments transformed to thorax. This is a change which progressed slowly from the head in a posterior direction. After the head had formed, the most anterior old abdominal segment transformed to the first thoracic segment, and the next behind became the second thoracic. As old abdominal segments were converted to thoracic in front, new abdominal segments arose at the rear. Thus a new worm with all its parts was reconstituted from part of an old worm.

Regeneration in many worms is like that described for *Sabella*. The first step is always the transformation of the most anterior remaining level to the far anterior part of a worm. All levels seem to contain the information for head, and many can use that information if they are at the anterior end of a piece of worm. Following the formation of head structures, the levels behind—following the rule that they can become the most anterior structures not present in front of them—transform to thorax. If differentiation is accomplished by the first region, inhibiting like differentiation in the second region and thereby forcing it to use the next part of its genetic program, there must be some way to move the repressors in an anteroposterior direction from cell to cell.

There is evidence that nerves are the polarizing agents during regeneration, and there is also reason to believe that they may be involved in transportation. Wherever nerves are found in animals capable of regeneration, their removal prevents regeneration. One example comes from the work of Wesley Coe (1930) on nemertean worms. Small pieces of worm were cut so that some included nerve cord and some did not. Those with nerve cord regenerated, and those without consistently failed to regenerate.

T. H. Morgan (1902) working with earthworms, learned that only when the ventral nerve cord extended all of the way to the anterior surface could a head regenerate from that surface. When he amputated the head and also removed a longitudinal strip of the antero-ventral part of the body including the anterior part of the ventral nerve cord, a head regenerated, but it was out of place. The head arose not from the most anterior remaining part of

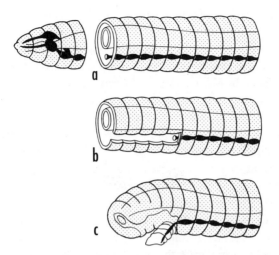

FIG. 6-7.   REGENERATION OF HEAD IN EARTHWORM FROM THE REGION AT THE
ANTERIOR END OF THE NERVE CORD [*redrawn from T. H. Morgan, 1902*].

If the anterior part of an earthworm is removed (*a*), and a strip of the body wall
plus the nerve cord is removed (*b*), the new head forms not at the most anterior
level of the worm but at the most anterior level where nerve cord is present (*c*).

the worm, but rather from that part of the body at the anterior terminus of
the shortened nerve cord (Fig. 6-7). There seems to be a general relationship here. In the Amphibia it is also where a nerve ends that tissues begin
to transform to the most distal structures of their area.

With respect to the Morgan experiment, one would never expect a
region with so much worm in front of it to become the anterior pole of a
developmental sequence. Apparently it is the most anterior part of the
nerve cord which determines which spot will transform to the missing
anterior part of the worm. Without a nerve cord in the anterior part of
a worm, the anterior part is unable to transform to head and unable to prevent a region with nerve cord behind it from transforming to head. It
would appear that the controlling information does not flow in a nerveless
region.

There was an interesting controversy over this work. A. J. Goldfarb
(1909) repeated the work and learned that he could get regeneration of head
from the most anterior part of an earthworm after the anterior part of the
ventral nerve cord had been removed. It was years later that Georg
Siegmund (1928) restudied the problem and resolved the difference. It
turned out that if the decapitated worms with the anterior part of the nerve

cord removed were kept long enough, they did regenerate heads from the anterior surface. However, it was only after a brain had regenerated that a head blastema appeared. It always appeared on the dorsal part of the anterior surface, in association with nerve fibers from the new brain. Ordinarily, when a head regenerates, it forms around the end of the ventral nerve cord on the ventral anterior surface. Whether heads regenerated in association with ventral nerve cord or with new brain, the nerves determined the very spot on the anterior surface which was to become the most anterior part of the regenerate.

In the annelid, *Myxicola aesthetica*, Y. K. Okada (1934) observed that simply cutting the ventral nerve cord without removing any structures was sufficient to cause the formation of a new head near the position of the cut.

More of the polarizing effect of the nerve cord was learned by L. P. Sayles (1940a,b, 1942) from studies on *Clymenella torquata*. He transplanted strips of ventral nerve cord with attached body wall under the dorsal surfaces of other worms. From the nearby dorsal surface a supernumerary head or tail could arise. In general, just as with the Amphibia, it was the surrounding tissues which had the greater say in determining the nature of the outgrowth. For example, in the posterior part of the body an outgrowth always became a tail no matter whether the piece of nerve cord came from the posterior, the middle, or the anterior part of the body.

If a tail starts growing from the back, it is quite clear that the direction of developmental control has been changed. Nerve cords terminating at the surface produce an end of a developing system. Whether that end will be a head or tail depends upon the general tissue neighborhood. Furthermore, the nerve cord is itself polarized. For an anterior piece of cord to operate, it must have its anterior end facing the surface. Pieces of nerve cord from the posterior part of the body set up new developmental axes only when their posterior ends face the surface. Pieces of cord from the middle of the body are doubly polarized and can induce outgrowths when either end is facing the body surface.

The nerve transplant in a worm or the deviated nerve in a salamander certainly induce developmental axes. What the blastema will become depends upon what is transforming. The posterior part of *Clymenella* is limited. It can only transform to the distal end of a tail. The amphibian limb is also limited. It can transform only to the distal end of a limb, never to a head or a tail. This does not mean that the information for other parts of the body has been lost in posterior parts. What one observes is that rather than reverting to a completely undifferentiated state, the bud forms the most distal structures of the particular region.

Two things are now clear. Nerves can establish the end of an axis along which differentiation occurs, and there are regionally specific substances which could act as agents of differentiation if they moved along the axis. Could nerves be involved in establishing a transport system for the repressors?

Here we turn to a different kind of study in an attempt to learn how nerves may make regeneration possible and how thay may determine its direction. Beginning with the studies of Lund on hydroids in the 1920s (Chapter Seven), it became clear that both the place where transformation would occur in a totipotent isolate and the direction of a differentiating axis could be determined by an applied electric field.

This type of study was extended to worms by Gordon Marsh and H. W. Beams (1952). They embedded mid-body isolates of planaria in agar to immobilize them and placed them in weak electrical fields for several days. The pieces regenerated normally when facing the negative pole over a range of field strengths. When facing the positive pole in the weakest field— less than 16.5 microamperes per mm^2 and with a voltage drop of approximately 80 millivolts per mm—there was no change in polarity. Heads regenerated normally from anterior surfaces in these very weak fields. However, as field strength was increased slightly, morphogenetic polarity began to change. Heads began to form at the original posterior surfaces as well as at the anterior surfaces, but these "posterior" heads did not persist. With slightly higher currents and potential differences, both heads persisted. Finally, with current densities in the neighborhood of 21-22μa per mm^2 and a drop of 225 mv per mm, there was complete and permanent reversal of polarity. Heads formed at former posterior ends and tails at former anterior ends.

An early step in the reversal of polarity appears to be a change in the direction in which messages are transmitted over the nerves. Even without a head, a planarian shows polarized neural mediated behavior. The anterior end moves in a characteristic way, annd there is rapid withdrawal of the anterior end from noxious stimuli. In the former posterior regions of worms whose polarity is being changed electrically, anterior behavior appears before anterior structures such as brain and eyes appear there. Since we already know that nerves determine the position of the distal end of an axis of regeneration. we are led to consider the possibility that an electrical field may change the direction of regeneration by first changing neural polarity. This then might result in the reversal of the direction of transport of the controlling agents.

There is another clue that a change in direction of neural activity is an early step in a change in morphogenesis. Berrill and Mees (1936), during their studies on *Sabella*, noticed that a change in direction of ciliary beat preceded loss of abdominal appendages, the first visible step in the transformation of abdominal to thoracic segments. This is another case in which the direction of neurally controlled activity changed before a change in morphogenesis was observed.

There is also an indication that without an operating polarized system, regeneration cannot occur. A piece of *Sabella* which did not show the usual wave-like progression of ciliary beat was isolated, and this isolate did not regenerate. It was recut later, and it behaved properly with successive waves of ciliary beats spreading across the surface. It also regenerated. These several clues lead to the suggestion that morphogenesis, always a polarized process, is dependent upon polarized neural activity. We also know that there is morphogenesis in embryos before nerves are formed. Whatever nerves do can apparently be done in another way in embryos. We shall return to this problem.

There are studies on earthworms indicating related involvement of electrical potentials and regional control of growth. Gairdner Moment (1946, 1949, 1953) removed different numbers of posterior segments of earthworms. The worms produced new segments posteriorly, until the total of old plus new segments was approximately 90. If there were 30 old segments, approximately 60 new ones would form. If 50 old ones remained, approximately 40 regenerated. It appeared that some kind of counting device was involved. Moment reasoned that it might be an electrical system. There were several reasons for suggesting this. He had been able to demonstrate potential differences along the worms as is the case in all organisms.

The first new observation was that as voltage was measured across more and more segments, it increased. During regeneration, when a certain voltage was reached no new segments were added. It seemed that each segment was acting as a battery. Even when quite small, the segments achieved full voltage, as do manufactured batteries. It seemed that the voltages of these segments, both old and new, might be adding up to an inhibitory voltage which prevented further segment formation.

The results of another experiment require that an amendment be made to Moment's suggestion. Sears Crowell (1958) cut earthworms into two pieces of unequal length. A long anterior piece was grafted to a long posterior piece, making a worm with too many segments. Shorter anterior and posterior parts were joined to make shorter than normal worms. Crowell was trying to learn whether the worms with too many segments— that is, too many if the inhibitory voltage suggestion was correct—could regenerate. Also, would the short worms make up their deficit by regenerating a greater than usual number of posterior segments after some were removed? It turned out that in 25 of 26 cases, after different numbers of caudal segments were removed, both the shortened and the lengthened worms followed the rule that the number of segments regenerated was just under the number removed. This means that the number of segments remaining does not determine how many will regenerate. Instead, a particular region is able to produce everything—or in this case almost everything—normally lying posterior to it. Wherever one finds regeneration, one learns that the regionality of the tissue at the regenerating surface, not the amount or kind of tissue behind it, determines the type and amount of outgrowth. Addition of new parts continues until all of the qualitatively different regions are present.

The fact that each level can support a certain amount of growth distal to it was seen in grafting studies on planaria by N.N. Schewtchenko (1936a and b). Bits of head tissue were grafted at different levels of the body of host worms. These induced outgrowths which became anterior parts. How much of the anterior part of a worm regenerated at the different body levels varied greatly. There was a regular order. Only heads appeared in the induced outgrowth from the anterior part of the body. The outgrowths were longer from mid-body levels and represented a greater part of the anterior part of the body. Outgrowths from the tail regions were longest and were almost complete worms.

In conformity with the general body of knowledge concerning regeneration this seems to be, first, a case of the grafts determining new axes of control. Then each body level would be free to transform to the most anterior structure not present along the new axis. If there were five regions, ABCDE, an outgrowth from B could transform to A, an outgrowth from C could transform to AB and an outgrowth from further back at D could form ABC. The outgrowths from anterior regions would be shorter only because the number of more anterior regions to which they could transform is smaller. The process of transformation should stop once all levels were represented—in this case when the outgrowth contained all levels anterior to the level of the region at the base of the outgrowth. Actually this is a description rather than an explanation.

The theory of inhibitory voltage or any theory which explains growth limitation or morphogenesis without consideration of the fact that the kind and amount of growth depends upon the region at its immediate base is faulty. However, we cannot escape the fact of bioelectric fields and that applied fields can determine the position and direction of regeneration.

Moment's work is still very important because it demonstrates that as the number of units in a line increases, so does the voltage. This may fit with the fact that a short non-regenerating piece of tissue may be made to regenerate by increasing the length of tissue behind it. What seems to be needed are studies in which both electrical potentials and regionally specific repressors are considered. Stephen Smith (1963) has made a start by showing that the regionally specific inhibitors of *Clymenella* can be moved in an electric field through sea water at pH8.1 toward the negative pole. Pieces of potential regenerates in the proper position downstream from the entrance point of the inhibitors in the field are specifically inhibited.

The relationship between electric fields and regional control has been studied more completely in hydroids (Chapter Seven).

References Cited

Benzer, S. 1962. The fine structure of the gene. *Sci. American* 206:70–84.

— Berrill, N. J. and Mees, D. 1936. Reorganization and regeneration in *Sabella*. 1. Nature of gradient, summation and posterior reorganization. *J. Exptl. Zool.* 73:67–83.

Betchaku, T. 1967. Isolation of planarian neoblasts and their behavior *in vitro* with some aspects of the formation of regeneration blastema. *J. Exptl. Zool.* 164:407–434.

Brøndsted, H. V. 1954. The time-graded regeneration field in planarians and some of its cytophysiological implications. *Proc. 7th Symp. Colston Res. Soc.* Colston Papers 7:121–138.

———. 1955. Planarian regeneration. *Biol. Rev. Cambridge Phil. Soc.* 30:65–126.

Child, C. M. 1941. *Patterns and Problems of Development*. Chicago, Illinois: University of Chicago Press.

Child, C. M. and Watanabe, Y. 1935. The head frequency gradient in *Euplanaria dorotocephala*. *Physiol. Zool.* 8:1–40.

Coe, W. 1930. Regeneration in nemerteans. II Regeneration of small sections of the body split or partially split longitudinally. *J. Exptl. Zool.* 57:109–144.

Crick, F. H. C. 1962. The genetic code. *Sci. American* 207:66–74.

Crowell, Sears. 1958. Tail regeneration in lengthened and shortened earthworms. *Biol. Bull.* 115:321.

Dubois, F. 1949. Contribution a l'étude de la migration des cellules de régénération chez les Planaires dulcicoles. *Bull. Biol. France Belg.* 83:213–283.

Flickinger, R. A. 1964. Isotopic evidence for local origin of blastema cells in regenerating planaria. *Exptl. Cell Res.* 34:403–406.

Goldfarb, A. J. 1909. The influence of the nervous system in regeneration. *J. Exptl. Zool.* 7:643–722.

— Hay, E. D. 1968. Fine structure and origin of regeneration cells in Planaria. *Anat. Rec.* 160:363.

Le Moigne, A. 1966. Étude du développement embryonnaire et recherches sur les cellules de régénération chez l'embryon de la Planaire *Polycelis nigra* (Turbellaire, Triclade). *J. Embryol. Exptl. Morph.* 15:39–60.

Lender, Th. 1956. L'inhibition de la régénération du cerveau des Planaires Polycelis nigra (Ehrb.) et Dugesia lugubris (O. Schm.) en présence de têtes ou de queues. *Bull. Soc. Zool.* 81:192–199.

———. 1960. L'inhibition spécifique de la différenciation du cerveau des Planaires d'eau douce en régénération. *J. Embryol. Exptl. Morph.* 8:291–301.

———. 1965. La régénération des planaires. In *Regeneration in Animals* ed. V. Kiortsis and H. A. L. Trampusch. Amsterdam: North-Holland Publ. Co.

Margolin, P. 1967. Genetic polarity in bacteria. *Amer. Nat.* 101:301–312.

Marsh, G. and Beams, H. W. 1952. Electrical control of morphogenesis in regenerating *Dugesia tigrina*. I. Relation of axial polarity to field strength. *J. Cell. Comp. Physiol.* 39:191–213.

Martinek, J. 1968. Project in Developmental Biology, Tulane University.

Miller, J. A. 1938. Studies on heteroplastic transplantation in Triclads. I. Cephalic grafts between *Euplanaria dorotocephala* and *E. tigrina*. *Physiol. Zool.* 11:214–247.

Moment, G. B. 1946. A study of growth limitation in earthworms. *J. Exptl. Zool.* 103:487–506.

———. 1949. On the relation between growth in length, the formation of new segments, and electric potential in an earthworm. *J. Exptl. Zool.* 112:1–12.

———. 1953. A theory of growth limitation. *Amer. Nat.* 87:139–153.

Morgan, T. H. 1902. Experimental studies of the internal factors of regeneration in the earthworm. *Arch. Entw-mech. Org.* 14:562–591.

Okada, Y. K. 1934. Régénération de la tete de *Myxicola aesthetica* (Clap.) *Bull. Biol. France Belg.* 68:340–381.

Okada, Y. K. and Sugino, H. 1937. Transplantation experiments in *Planaria gonocephala* Duges. *Jap. J. Zool.* 7:373–439.

Pedersen, K. J. 1961. Studies on the nature of planaria connective tissue. *Zeit. f. Zellforsch.* 53:569–608.

Pentz, S. and Seilern-Aspang, F. 1961. Die Entstehung des Augensmusters bei *Polycelis nigra* durch Wechselwirkung zwischen dem Augenhemmfeld und der Augeninduktion durch das Gehirn. *Arch. Entw-mech. Org.* 153:75–92.

Reutter, K. 1967. Untersuchungen zur ungeschlechtlichen Fortpflanzung und zum Regenerationsvermögen von Lineus sanguineus Rathke (Nemertini). *Arch. Entw-mech. Org.* 159:141–202.

Santos, F. V. 1929. Studies on transplantation in Planaria. *Biol. Bull.* 57:188–199.

———. 1931. Studies on transplantation in Planaria. *Physiol. Zool.* 4:111–164.

Sayles, L. P. 1940a. Buds induced by implants of anterior nerve cord and neighboring tissues inserted at various levels in *Clymenella torquata*. *Biol. Bull.* 78:298–311.

———. 1940b. Buds induced by implants of posterior nerve cord and neighboring tissues inserted at various levels in *Clymenella torquata*. *Biol. Bull.* 78:375–387.

———. 1942. Buds induced in *Clymenella torquata* by implants of nerve cord and neighboring tissues derived from the midbody region of worms of the same species. *Biol. Bull.* 82:154–160.

Schewtchenko, N. N. 1936a. Polaritätsumkehr bei der Unterdrückung der dominierenden Region bei Planaria. *Zool. Anz.* 115:232–44.

———. 1936b. Der Einfluss des Korpteiles von hoher physiologischer Aktivität auf den Regenerationsvorgang bei Planarier. *Proc. Inst. Zool. Kiev.* 43–64.

Siegmund, G. 1928. Die Bedeutung des Nervensystems bei der Regeneration, untersucht an Eisenia. *Biol. Gen.* 4:337–350.

Smith, S. D. 1963. Specific inhibition of regeneration in *Clymenella torquata*. *Biol. Bull.* Woods Hole 125:542–555.

———. 1964. Time-action relationships of the specific inhibitor of caudal regeneration in *Clymenella torquata*. *Proc. Soc. Exptl. Biol. Med.* 116:83–85.

Stephan-Dubois, F. 1965. Les néoblastes dans la régénération chez les planaires. In *Regeneration in Animals and Related Problems* ed. V. Kiortsis and H. A. L. Trampusch, pp. 112–130. Amsterdam: North Holland Publ. Co.

Teshirogi, W. 1963. Transplantation experiments of two short pieces of a freshwater planarian. *Bdellocephala brunnea. Jap. J. Zool.* 14:21–48.

Tucker, M. 1959. Inhibitory control of regeneration in nemertean worms. *J. Morphol.* 105:569–599.

Wolff, E. 1962. Recent researches on the regeneration of Planaria. In *Regeneration* ed. D. Rudnick. New York: Ronald.

Wolff, E., Lender, Th., and Ziller-Sengel, C. 1964. Le rôle des facteurs autoinhibiteurs dans la régénération des Planaires. *Rev. Suisse Zool.* 71:75–98.

Woodruff, L. S. and Burnett, A. L. 1965. The origin of the blastema cells in *Dugesia tigrina*. *Exptl. Cell Res.* 38:295–305.

Ziller-Sengel, C. 1967. Recherches sur l'inhibition de la régénération du pharynx chez les planaires. I. Mise en évidence d'un facteur, auto-inhibiteur de la régénération du pharynx. *J. Embryol. Exptl. Morph.* 18:91–105.

CHAPTER SEVEN

Regeneration in Coelenterates

The major question in the study of regeneration is how cells communicate with each other and what they say to effect the reconstitution of an integrated whole after removal of a part. Studies of the problem in a variety of organisms have been given in the earlier chapters. It appears from these studies to be generally true that regeneration begins when repressive control is removed and continues until a complete series of repressions is reconstituted. The coelenterates have been very useful in studies of reconstitution. Some aspects of the problem seem to be better understood in this phylum than elsewhere. The perennial sub-problem of which cells take part in regeneration has been most thoroughly explored in this group and will be considered first.

We go back to the last part of the nineteenth and the early part of the twentieth century to learn why people began to believe that there are special totipotent reserve cells involved in regeneration and budding.

Carl Chun (1896) described a group of jellyfish collected on the Atlantis expedition. These jellyfish—or medusae—were unusual in that they not only reproduced sexually with eggs and sperm, the usual way for medusae, but also reproduced by asexual budding. Chun described buds arising in ectoderm. They became two-layered with the outer layer remaining ectoderm. Central cells in the ectodermal bud regrouped as an inner epithelium. This layer became the new endoderm. Subsequent outgrowths and foldings of both layers led to the production of new small medusae.

Later F. Braem (1908) reworked Chun's specimens and introduced an imaginary cell, one which he did not see, to explain the process of budding. He believed that there must have been members of a special line of undifferentiated cells which migrated to the site of budding. To understand why Braem could not accept Chun's description and felt the need for implicating an unseen cell type, one has to understand the nature of a revolution in biological thinking which was occurring at the time.

August Weismann (1892), and others who followed him in the next few decades, had put across the notion that differentiation is essentially a process of segregation. According to this view, a fertilized egg contains all of the determiners for the many parts of the future body. With each cell division some of the determiners pass to one daughter cell and others to the other daughter cell. Materials determining head are supposed to end up in cells forming head and not in cells forming tail. If one accepted this scheme, and most students of animal development did, one could not accept the observation that differentiated cells of one type or one region could, during the course of regeneration, become cells of another type or region. They simply did not have the information for anything except what they already were. The fact of regeneration in adult animals required the invention of a line of cells like germ cells with a full set of determiners which could differentiate into a variety of types. These theoretically necessary cells came to be known as reserve cells.

In time, certain real cells which could be distinguished under the microscope were thought to be reserve cells. The neoblasts of worms and the interstitial cells (I-cells) of coelenterates came to be thought of as reserve cells. Long after differentiation by segregation was disproved (Chapter Eight) the reserve cell, invented to support it, lingered on. An appreciable amount of information incompatible with the theory of regeneration by reserve cells has accumulated over the years. Only the information gained from studies of regeneration in coelenterates is covered here.

The first type of evidence was derived from histological studies of stages in the transformation of one structure to another. N. M. Stevens (1902) and Emil Godlewski (1904) described the transformation of pieces of stem to hydranths in the marine hydroid. *Tubularia* (Fig. 7-1). The stem is a two-layered hollow cylinder. Part or all of an isolated stem, depending on its size, transforms to a hydranth. One feature of this transformation is the formation of two series of longitudinal ridges along the stem cylinder. These ridges arise from typical ectodermal and endodermal cells with their usual characteristic inclusions. The ridges delaminate from the rest of the cylinder except at their bases and are then new tentacles. This description is incompatible with regeneration by reserve cells. What was observed was a step-by-step transformation of an entire region of a stem to a hydranth.

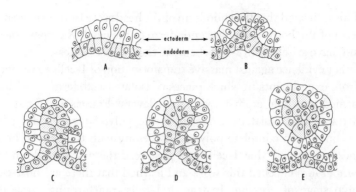

FIG. 7-1.   REGENERATION OF A TENTACLE IN TUBULARIA [*adapted from N. M. Stevens (1902)*].

a: A part of the cylindrical coenosarc of a stem.

b: Both ectoderm and endoderm hump up into a ridge of which this is a cross section.

c, d: Stages in the enlargement of the tentacle ridge.

e: Ectoderm meeting ectoderm and undercutting the ridge. In time the tentacle is cut free from the underlying coenosarc except at its point of attachment at its base.

If reserve cells formed the new hydranth, they would have to replace the ordinary cells of the stem and then produce a hydranth. There is no replacement of regular stem tissue prior to the transformation of the whole tissue to hydranth. If the relatively few interstitial cells originally present were to produce the whole regenerate, they would have to divide very rapidly. The entire time for transformation from stem to hydranth can occur in one day. There is not time enough for the few interstitial cells to increase to the amount of tissue which becomes the hydranth. Furthermore, the transformation is accomplished without or with insufficient cell division (Stevens 1902; Godlewski, 1904; Tardent, 1965). Appreciable cell division comes later as the new hydranth begins to grow.

J. Kanajew (1930) using *Hydra* demonstrated in another way that there is no massive transformation of I-cells to new head during regeneration. He had placed his hydras in a solution of methylene blue. Eventually all of the dye was to be found in the endodermal cells. When the anterior end of a body with blue endoderm became a head, throughout the transformation, from the time it was body until it became hydranth, the endoderm remained blue. This indicated that the endoderm of the head arose by transformation of endoderm of the body. The blue endoderm of the head, Kanajew reasoned, could not have arisen from I-cells because they were colorless.

Further evidence against massive transformation of I-cells to regenerates arose from experiments in which pieces of isolated ectoderm or endoderm transformed to whole animals with both layers. Francis Gilchrist (1937) demonstrated that isolated ectoderm of the polyp stage of the jellyfish *Aurelia* could form complete polyps. In keeping with the general belief of the time, he assumed that I-cells had produced the new endoderm. Sonia Steinberg (1963) checked this work and learned that there are no I-cells in the polyp stage of *Aurelia*. Instead of I-cells transforming, some of the ectoderm transformed to endoderm. Similarly Edgar Zwilling (1963) demonstrated that the hydroid *Cordylophora* can reconstitute whole polyps from ectoderm alone. *Hydra* too can be reconstituted from one layer, in this case from the endoderm (Diane Normandin, 1960; Julian Haynes and Allison Burnett, 1963). Haynes and Burnett used *Hydra viridis*. The endodermal cells of this species harbor a green symbiont. They were able to follow endodermal cells as they lost their symbionts and became colorless ectoderm.

The above several studies indicate that the I-cells are not the sole transformers. In addition we know from the Steinberg experiment with *Aurelia* that no I-cells need be present. There are other experiments which indicate that no I-cells need be present during reconstitution in species which usually contain them. This knowledge is derived from the fact that regeneration can occur in regions naturally devoid of I-cells or after they have been eliminated by special treatments. Ordinarily all parts of a hydra except tentacles and the base of the stalk near the posterior end of the animal can reconstitute the whole organism. These non-regenerating regions are without I-cells and the correlation was originally thought to be a causal one. It was quite generally accepted that the base of the stalk could not regenerate because it lacked I-cells. B. P. Tokin and G. P. Gorbunowa (1934) and L. von Bertalanffy and M. Rella (1941) cut up I-cell free portions of the stalks in such a way that the old organization was disturbed. The pieces later reorganized into whole hydras. Everything including I-cells formed after disruption of the old organization. This indicated that rather than the I-cell being the progenitor of the various somatic cell types, it could arise from at least one of them.

In addition, reconstitution of whole hydras from parts follows the eradication of all I-cells by sufficient doses of X-rays (P. Brien and M. Reniers-Decoen, 1955). The hydras do not feed or grow after these treatments and are destined to die in a few days and all I-cells disappear, but still parts can reconstitute wholes. The same is true after eradication of all I-cells by nitrogen mustard (Fred Diehl and Allison Burnett, 1965).

The several kinds of evidence against reserve cells amount to a proof that there is no such special cell for regeneration. Further discussion of this problem with additional cytological information has been given by Elizabeth Hay (1966).

The major function of the I-cell is transformation to cnidoblasts, the stinging cells (D. B. Slautterback and D. W. Fawcett, 1959). PierreTardent (1965) has made the point that although there is considerable evidence that the I-cell is not the sole cellular source of material for regeneration, it may also take part in reconstitution by transforming to other cell types.

P. Brien and M. Renier-Decoen (1949) demonstrated that cellular transformation is a constant feature of life in *Hydra*. They demonstrated this by staining a ring of cells below the tentacles. In a few days the ring of stained cells had apparently moved down the trunk to the base of the stalk where it disappeared as the cells died.

Originally it was thought that there was a growth zone near the base of the tentacles creating a population pressure which pushed older cells down the column. When Richard Campbell (1967 a, b) studied the distribution of mitotic figures and the uptake of tritiated thymidine into DNA, he learned that there are cells all over the body preparing to divide and dividing. Any cell in the trunk gets closer and closer to the base as cells die at the base. This would happen whether cells were dividing above them on the column or not. They are getting closer to the base because the base is constantly removed by death and there is not sufficient growth between the base and the stained ring to prevent the ring from eventually becoming the base.

As the stained ring came to lie closer and closer to the base, it occupied different parts of the body. When a cell was stained behind the tentacles it was a trunk cell. Later, if that cell happened to be in the proper position when spermaries or ovaries arose, it became part of one of those organs. If a bud began to arise when the stained ring was at the top of the budding region, only the upper part of the bud developed from stained cells. When a ring had apparently moved to a slightly more posterior level when budding began, just the lower part of the bud was stained. Eventually the cells in the remnant of the ring underwent the final differentiation which in this case was death.

Each cell, as it came to lie closer and closer to the base, assumed the characteristics proper for that region. Constant natural death of basal cells causes constant regional transformation, just as removal of cells with a knife does when regeneration is artificially initiated.

The organization of the the *Hydra* is maintained, although the membership of the region constantly changes. The stained band does not really move to the base. Instead, as basal death proceeds, the base comes closer to the ring. The ring does not move through region after region. It becomes region after region always following the rule that it will produce the most distal structures not present distal to it.

Now knowing that type of differentiation depends upon position, we turn to the major problem—how do cells communicate and what do they say with the result that they differentiate?

The analysis of regeneration began with Abraham Trembley's studies of Hydra published in 1744 and reviewed by John Baker (1952). After the initial work in the eighteenth century it was learned that many animals can regenerate, but little was learned about the process until the late nineteenth and the early twentieth centuries. At the time when so much was being learned from plants (Chapter Four), some very able experimental biologists, Charles Manning Child, Hans Driesch, Jacques Loeb, and Thomas Hunt Morgan turned to the coelenterates and learned that regeneration is basically similar in plants and in coelenterates. Much of the work was concentrated on Hydra and the marine hydroid, *Tubularia* (rev. by Barth, 1940; Child, 1941; Tardent. 1963). *Tubularia* became a favorite for study because of its relatively large size and because it regenerated so rapidly. If a hydranth is removed from a stem, there is usually a primordium of a new hydranth visible in the stem within a day, and by two days a new small functional hydranth can emerge from the perisarc, a secreted casing around the stem.

One thing that became clear from the early studies was that a portion of a stem which was transforming to a hydranth could prevent other nearby regions from doing the same. This kind of control came to be known as dominance and has been demonstrated in many organisms (Child, 1929).

Herbert W. Rand, J. V. Bovard, and D. E. Minnich realized by 1926 that they could establish a new center of control of organization by grafting a head into the side of a whole hydra. This could also be done by grafting only a part of the head, a piece from the mouth region. Ethel Browne (1909) had first observed that hypostomal pieces which made part of a head when grafted into a trunk also induced a new axis of growth. The adjacent trunk transformed to a head—that is, all of the head except what the graft made. Behind the head in order arose the rest of the hydra except the level posterior to the level of the trunk where the graft had been placed. The result was a Y; the original head and the one resulting from the action of the graft were at the tops of the arms of the Y. The new out-growth became as long as that part of the original hydra anterior to the level of the graft. The two arms of the Y became the same length. Rand and his students cut off the original arm of the Y in a number of cases. If the stump was short, it did not regenerate a new head. Heads would regenerate on both stumps if both heads were removed. It was realized that a head controlled the organization behind it by preventing head formation, but it did not inhibit non-head development.

Five to ten mm-long stem isolates of *Tubularia* usually produce only one hydranth. Under normal conditions the new hydranth arises at the distal end of the stem, the end away from the base of the stem and closer to the top. This is the same end at which the original hydranth had grown.

The distance over which dominance extends varies considerably. In the spring and fall at Woods Hole, Massachusetts, when the colonies of *Tubularia crocea* are growing rapidly, the distal end of an isolate may prevent a proximal end 15-20mm away from forming a hydranth. If they have been in warmer water, approaching 21°C., they are about to become dormant for the summer (John Moore, 1939). Just before they become dormant, regeneration is relatively slow and the distance over which dominance is exerted is much less. Sometimes one finds hydranths arising at both ends of shorter stems, 5–10 mm long. A few days later in the summer stems isolated from freshly collected colonies fail to regenerate.

The early workers learned that when regeneration was prevented at the distal end, the proximal end was no longer dominated. Loeb pushed distal ends into sand. This prevented transformation at that end and thereby freed the other end to transform. The same result was accomplished by Morgan by tying a ligature around the distal end. The ligature pursed the perisarc, thus closing the opening between tissue and sea water. The closed end failed to regenerate, and the still open proximal end became the regenerating region. In similar fashion, W. B. McCallum (1905) had prevented growing tips of plants from dominating more proximal buds when he surrounded the tips with N_2. These treatments do not kill but they do suppress some of the metabolism.

Several questions which have now been answered or are being answered arise from this early work. Why is the regeneration from the end or ends of the stem? Why is it almost always the distal end which regenerates when only one does? How can one region dominate another far away without suppressing its own transformation? Finally, what passes between regions resulting in dominance?

The first question about why regeneration occurs at an end was approached with an experiment by Edgar Zwilling (1939). He suspected that the opening through the perisarc at the end might be important because it was facilitating the passage of some substance, possibly oxygen. What he did was to tie off the two ends of the stem and make a new opening in the side. A piece of perisarc was removed as carefully as possible to minimize injury to the underlying tissue. As can be seen (Fig. 7-2), the lateral openings became centers around which stem transformed to hydranth. Instead of the tentacle ridges being oriented parallel to the main axis of the

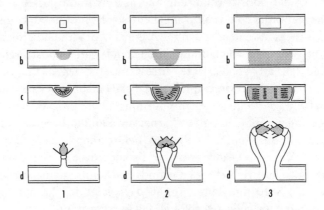

FIG. 7-2.   THE DEMONSTRATION THAT REGENERATION OCCURS IN TUBULARIA IN THAT REGION WITH CLOSEST CONTACT WITH THE ENVIRONMENT [*adapted from Edgar Zwilling (1939)*].

The two ends of the isolated pieces of stem were sealed with ligatures. The only free contact with the sea water was through a hole cut in the non-living perisarc. Beneath the hole the living tissue was bathed by the sea water.

In the cases where the hole was small (Series 1), the primordium was small and the proximal tentacle ridges pointed toward the opening. A single complete hydranth emerged (d 1).

In Series 2 the opening was larger and most of the tentacle ridges pointed toward the opening. The hydranth which emerged was single distally but doubled proximally.

In Series 3, with the largest opening, a much greater part of the stem transformed to hydranth. When distal ridges appeared, there were two distinct sets and two complete, although joined, hydranths emerged.

This work demonstrated that contact with the environment determined position, size, and polarity of the regenerates.

stem, they pointed toward the lateral opening. The polarity of the system was determined by this new door to the environment.

Could it be that something entering or leaving through that door was favoring the nearby tissue so that it could transform to hydranth? The stem is totipotent in that all levels can transform to hydranth. The opening was certainly determining the level.

At the time of the lateral door experiment, there was also much to indicate that a regenerating region had the highest general metabolism of all nearby regions. Child and his associates had shown in many organisms that regenerating regions and especially anterior ends of regenerating regions have the ability to oxidize and reduce dyes more rapidly than neighboring regions. In many cases the ability to metabolize rapidly dropped off as a gradient. These regions are also most easily poisoned by a variety of poisons (summarized by Child, 1941).

James A. Miller, Jr. (1937), one of Child's students, had gone on to show in *Tubularia* that the relationship between highest relative metabolism and site of regeneration is a causal one. This was demonstrated by another simple experiment. The experimental question was, "If transformation from stem to hydranth occurs at distal ends because metabolism is highest there, would it occur at proximal ends if their metabolism were made greater?" Miller constructed a two-compartment chamber with a partition between the two compartments. In the partition were many holes, and into these were inserted stems of *Tubularia* with the two ends in different compartments. When oxygenated sea water passed through one compartment and non-oxygenated through the other, the ends bathed by the oxygenated water transformed to hydranths. It did not matter whether they were distal or proximal. The other ends in the non-oxygenated water apparently received enough oxygen by way of the gastrovascular cavity to live, but they did not transform. Likewise warmer water—in the range tolerated by *Tubularia*—caused the ends in it to become hydranths, while the cooler ends remained stem. The anterior end of an axis of differentiation is the anterior end because of its relatively high metabolism.

This is something basic to our understanding of differentiation. The system is one in which all levels have the information, probably genetic, to make a hydranth. One region is given a higher rate of metabolism—this time by the experimenter. It uses certain genes and makes a hydranth. At the same time, and this is the feature preventing chaos, the development of one precludes the development of others nearby.

An important step toward understanding how dominance is achieved was taken by Lester Barth (1938a). Barth measured the rate of regeneration in isolates taken from different levels of the same stem (Fig. 7-3). There was a regular, gradual decrease in rate of regeneration of the isolates from distal to proximal. The distal end of the distalmost piece transformed more rapidly than all distal ends from lower stem levels. The distal end of the second piece transformed more rapidly than the distal end of the third piece, and so on to the base of the stem. In general, the regenerates from more distal were also larger. There was in inherent graded difference along the stem.

It was already known that the dominance of one end of a stem over the other could be blocked by tying a ligature between them. When isolated by the ligature, both could regenerate. Barth then compared the time for regeneration of just the distal and proximal ends. This was done by tying two ligatures back a ways from the ends (Fig. 7-3). The few millimeters of distal tissue transformed to hydranth faster than the few millimeters of proximal tissue. In similar stems without the ligatures, a hydranth formed at the distal end, but rarely did one arise at the proximal end. It appeared to be a case of the rich getting richer and the poor getting poorer. Whenever this situation—the faster region forming the structure—has been observed throughout the plant and animal kingdoms, the first thought has always been that the rich have some advantage enabling them to take from the poor. A competition for something in short supply always becomes the initial hypothesis.

Several things speak against control by competition for supplies in *Tubularia*. First, in an isolated stem of 5–10 mm, possibly a third or a fourth will be converted to hydranth. If such a length of stem is divided into 1.5mm pieces, all of the material may transform to hydranth. It cannot be a case of too little pre-hydranth material. It can all become hydranth.

Two other observations make it appear that the two ends are competing for materials. Many stems of *Tubularia* contain a reddish pigment. The first indication of which end is going to form the hydranth is the migration of the pigment toward the regenerating end and away from the non-regenerating end. In addition, cells have been shown to move away from proximal ends toward more distal regions. This has been proposed as the mechanism of dominance (Malcolm Steinberg, 1954, 1955). Steinberg studied summer stems. These contain relatively little living material, and the correlation between migration and dominance is good and would appear to be causal. However, if one studies winter stems, one learns that distal regions dominate proximal regions without taking away from their materials for regeneration. In the winter the stems have a thick living layer, the coenosarc, so thick that it almost obliterates the central cavity. These stems have enough

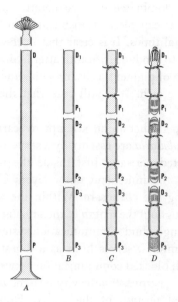

FIG. 7-3. GRADATION IN RATE OF REGENERATION FROM DISTAL TO PROXIMAL IN TUBULARIA [*adapted from Lester Barth (1938a)*].

a: A long piece of stem isolated at its distal end from the hydranth and at its proximal end from its base.

b: The stem cut into 3 pieces and the distal and proximal ends numbered from distal to proximal.

c: A few millimeters of stem at the ends of each piece isolated by tightly applied ligatures. This enabled Barth to learn the speed of regeneration at the various levels without competition or dominance with the other end or with the intervening stem.

d: One sees a gradation in rate from distal to proximal. D1 has a completed hydranth with distal and proximal tentacles. D2 is a little behind with proximal tentacles but only distal ridges. D3 is a little behind D2 with the distal ridges just beginning to appear. At the P ends, P1 is the most advanced but has only achieved the stage seen in D3. P2 is behind P1 and shows 2 bands with no distinct ridges as yet. P3 is in a very early stage showing only one band, the earliest recognizable primordium of a hydranth.

stored material so that they can regenerate several times. Proximal regions cut away after being dominated can regenerate when isolated. Even the original proximal tip is capable of transforming to hydranth after having been dominated several times. It is clear that in these winter stems which have enough material to produce hydranth after hydranth, dominance is not the result of depletion of pigment, cells, or some unknown ingredient. The rich can dominate the rich. It is still true that the faster dominate the slower.

Barth went a step farther in an attempt to learn the pathway of dominance. Stems of *Tubularia* are essentially a series of cylinders. The outer cylinder is the secreted perisarc. Just inside the perisarc are two more cylinders. These are the inner and outer layers of cells, ectoderm and endoderm, comprising the coenosarc. Within the endoderm is a central cavity extending throughout the length of the organism. This gastrovascular cavity serves as the single fluid communication system. The gastrovascular cavity, as its name implies, serves both as a digestive system and a circulation system. Barth blocked communication between distal and proximal regions by way of the gastrovascular cavity by injecting a drop of mineral oil which filled a central portion of the cavity. Dominance was blocked, indicating that the control of one end over the other was by way of the gastrovascular cavity. A glass rod in the cavity also blocked dominance (S. M. Rose and F. C. Rose, 1941). A short glass tube inserted in the gastrovascular cavity permitted some circulation through it, and there was some dominance. This seemed at the time to indicate that control passed by way of the fluid. However, it was learned from later studies that cells can migrate along the surfaces of the tubes and reestablish tissue connections.

Child in a personal communication expressed the belief that the materials which were placed in the gastrovascular cavity were also in contact with the endoderm and might have interfered with transmission of something through the coenosarc. There are several things now which indicate that control passes from cell to cell rather than through the cavity. The major part of this evidence is introduced later in this chapter, but one part is pertinent here.

James Miller (1959) isolated stems and maintained a constant flow of sea water through their gastrovascular cavities. The sea water was sucked in at the distal end and removed at the proximal end. These regenerated very well and only at the distal end of the stem. This demonstrated that the distal end could not have regenerated because it received anything from more proximal levels by way of the cavity, since the flow was away from the distal end. It also made it very difficult to believe that distal was dominating proximal by something passing from distal to proximal regions through the cavity. Although the direction of flow was from distal to proximal, the flow was great enough to have flushed out most distally produced materials before they could have passed into proximal cells. We shall return to the question of pathway of control.

It was established by the experiments outlined above that more rapidly regenerating regions can suppress regeneration in more slowly regenerating regions if they are connected and not too far apart. The fast region must suppress the slower region in some way other than by competition for something in short supply. An inkling as to how dominance was exerted came from an attempt to learn more about how oxygen might stimulate regeneration.

James Miller had already shown that one can determine position of regeneration and change polarity by giving a region a metabolic advantage. Lester Barth (1938b) had demonstrated that as O_2 tension is decreased, a level is reached where stems can maintain themselves but not regenerate. Also, the length of the stem over which transformation occurs increases with increasing O_2 tension. Finally, the Zwilling experiment had shown that any region where there was contact between sea water and coenosarc could transform. The question was, does regeneration occur where it does because that region is getting more oxygen?

With the knowledge that regeneration is blocked if both ends of a stem are ligated, the test was first made by injecting oxygen bubbles into the gastrovascular cavity and then ligating the ends. If it were really oxygen which a ligature was cutting off and thereby preventing regeneration, ligated stems with a good internal supply of oxygen should regenerate. The stems did start. One could see the red pigment beginning to concentrate, but regeneration did not proceed to the point of the formation of new structures. Why did it stop? New oxygen was added, but still the new hydranths did not form while the stems were ligated. It appeared that something besides oxygen was in short supply, or that an opening was needed for the release of something.

Another type of experiment indicated the same thing. If instead of ligating the end of a stem, we applied a loosely fitting glass cap, that too would prevent regeneration at the end which it covered. The addition of an oxygen bubble in the cap reversed the action of the loose-fitting cap. However, a more tightly fitting cap, even with oxygen, always prevented regeneration. It was becoming clear that something besides oxygen might be involved. It appeared that the loosely fitting cap was permitting the passage of something else.

At this point Florence Rose got the idea that regeneration was failing in the absence of an opening because something inhibitory to regeneration was being produced and not being released. Such a notion seemed to be required by the fact that the ligatured stems with injected oxygen could regenerate if the ligatures were cut away in time. The idea was that if self-inhibitors were involved, these should increase with time in ligatured stems. Stems ligatured for a short time should be less affected than those ligatured for a longer time.

The experiment was performed by tying off both ends of one group of stems. At the same time, other stems were isolated but not ligatured. At intervals ligatures were removed from some of the stems. It turned out that the transformation in the ligatured stems required very few hours more than in the non-ligatured stems if the ligatures were removed in 20 hours. If the ligatures remained for much longer than 20 hours, the stems gave no evidence of having started to regenerate while ligatured. It took them as long to regenerate as control stems freshly cut at the time the long-applied ligatures were removed. In time, something occurred in the ligatured stems which caused them to lose the initial steps which they had taken.

The fact that the stems had started to regenerate while both ends were tightly tied off indicated that the initiation of regeneration did not require a locally higher supply of oxygen or anything else. Also, the initiation of regeneration did not require the removal of anything from an open surface. The isolation of the stem from its hydranth alone was enough to cause the stem to start to transform to hydranth.

Under most laboratory conditions there is a failure of ligatured stems to complete regeneration because too little O_2 gets in or too much inhibitor material is retained. If ligatured stems are kept in running sea water, some can regenerate (Rose and Rose, 1941). Similarly Barth (1938b) had learned that if ligatured stems are shaken, they can regenerate. Apparently, if ligatured stems are lying in still water, there is insufficient exchange through the perisarc to enable them to complete transformation. However, even in still water regeneration began in ligatured stems. This indicates that no externally induced local advantage was necessary. Simple isolation of the stem was enough to cause it to transform to hydranth. There was still an inherent gradient in these ligatured stems. They almost invariably produced the new hydranths at the distal end.

Pierre Tardent (1964) has made it clear that there is a disto-proximal gradient of O_2 uptake paralleling the gradient in regeneration rate in *Tubularia*. However, regeneration does not involve an increase in O_2 uptake. Instead, there is a general drop during the early stages.

The several studies outlined above teach us that inherent differences grading off from distal to proximal are such that a distal region can regenerate faster. Anything an experimenter does to make another region faster enables it to become the transforming region and prevents its slower neighboring regions from doing the same. If this control is not by competition for something, could it be by inhibitors as hypothesized above?

Tubularia does produce an inhibitor or inhibitors of regeneration which can be collected in sea water (Rose and Rose, 1941; Tardent, 1955; rev. 1963; Tweedell, 1958a, b). Doubt was cast on the inhibitor as a normal agent of control because products from decaying *Tubularia* and other organisms can prevent regeneration in *Tubularia* (Chandler Fulton, 1959). However, it was demonstrated by Kenyon Tweedell (1958a and b) and by Pierre Tardent and Hermann Eymann (1959) that the inhibition can be produced by hydranths and not by stems when bacteria are kept to a minimum.

Natural inhibitors of regeneration have also been demonstrated for other genera of hydroids; *Cordylophora* (Penzlin 1957). *Campanularia*, *Coryne* and *Clava* (Beloussov and Geleg, 1960) and for Hydra itself (Ruffing, 1966). L. V. Bleoussov and Z. Geleg showed that the inhibitors can cross react among genera.

It has been learned not just from coelenterates but from studies on a variety of organisms that as organisms develop in water, they add to that water substances which inhibit the development of members of their own species (rev. by Rose, 1960; Rose and Rose, 1965). The inhibitory products are relatively specific. If eggs of one species of sea urchin are cultured in sea water, removed from the water, and a second batch of young developing eggs added, the second batch fails to develop. Eggs of a closely related species may live longer, but they also fail. The same water serves as an excellent culture medium for eggs of a more distantly related sea urchin of another family (H. M. Vernon, 1899). The same relationship was shown for *Daphnia* (V. H. Langhans, 1909). Similarly water-borne products of one species of frog tadpole are very effective in stopping development in the same species during the tadpole stage (Christina Richards, 1958) and during differentiation as early as the gastrula stage (Gwynn Akin, 1966). The effect of one species on another within the same genus is less and almost vanishes when even more distantly related animals are involved. For example, when salamander culture water is given to eggs of frogs (Akin, 1966), there is scarcely any effect.

Throughout the work with natural inhibitors of growth and differentiation it has been learned that more rapidly growing organisms can block the development of more slowly growing ones. A single tadpole produces a level of inhibition which will slow itself down. That same level of inhibition may prevent all growth in a sibling less advanced in development (Rose and Rose, 1961, 1965).

Returning now to inhibitors of regeneration, we find an example of this general rule in *Tubularia*. Pierre Tardent and Hermann Eymann (1959) and Tardent (1960) grafted hydranths and regenerating primordia to the proximal ends of regenerating stems. Adult grafts blocked all regeneration in recently isolated stems. When combinations were made so that the stage of development of graft and host differed, it was always the more advanced which continued to develop. The more advanced also suppressed development of the less advanced. When both had reached the same stage of development at the time when they were combined, both continued to develop. The system is such that self-inhibition is minimized. As a primordium develops, it releases something which prevents less advanced nearby regions from achieving the same stage of development.

Both in *Tubularia* (Tardent and Eymann, 1959) and in *Rana* tadpoles (Rose and Rose, 1961), the more advanced produce a greater effect on the earliest stages and the more advanced are less susceptible to the inhibition. Any individual of any stage produces a level of inhibition which it can tolerate. This same level of inhibition will slow down or stop a less advanced individual.

As Tardent and Eymann pointed out, these findings help one understand a series of relationships which had seemed paradoxical. In pieces of intermediate length, e.g. 8 mm, dominance of distal over proximal regions is usually complete. In longer pieces dominance decreases with increase in distance between the ends. What did not make sense was that dominance also decreased in short pieces and could disappear in very short pieces, under 2 mm. Hydranths formed at both ends of very short pieces. Often there was not enough material for two complete hydranths. In that case just distal parts of hydranths formed. Both members of these bipolar forms differentiate at the same rate. They are so close to each other along the gradient that neither has a sufficient head start to inhibit the other. Just as self-inhibition is minimal, so is the effect of two of the same stage on each other even when they are very close.

So far in this account of dominance in *Tubularia* we have been considering the interaction between regions resulting in the establishment of one or two axes of differentiation. Now we inquire how the regions along a single axis interact to produce the several different levels of a hydranth.

We start with the knowledge that all parts of the stem are totipotent.

Any part of a stem can become any part of a hydranth. When tiny pieces are isolated, they may fail to regenerate or they may make part of a hydranth. They always make the same part-an anterior part. The smallest regenerators form nothing more than an anteriormost region including the anterior tentacles (Fig. 7-4). With more material available, more posterior structures may form. Thus we learn that every part of the stem is potentially the anterior part of a hydranth. We also know that in larger isolates the posterior portions, which could have become anterior structures if left alone, are now relegated to the production of things not forming anterior to them. How is this done in such an orderly way that a perfect hydranth with all parts perfectly aligned can regenerate?

One can make combinations of regenerating parts and study their effect on each other (S. M. Rose, 1957). In order to learn from this method, one has to choose the proper time to make the graft combinations. That time is just before new structures appear. If taken much earlier, there is a tendency for all parts to dedifferentiate, lose all regionality, and start afresh. If regions are grafted after they have produced new structures, they do not

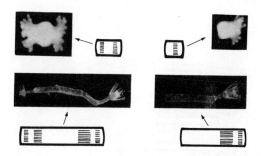

Fig. 7-4.   Regeneration of pieces of stem of Tubularia of different lengths
[*from* S. M. *Rose (1967)* Jour. Gerontology 22: II 29].

Lower right: A stem 10 mm long usually produces two sets of tentacle ridges at its distal end (diagram) and these form a single hydranth (photograph).

Lower left: If isolated stems are longer than 10 mm it is likely that the distal dominance will not be sufficiently effective to prevent regeneration at the proximal end.

Upper: If the isolates are too small to produce whole hydranths they may produce just the anterior part of a hydranth (right) or two anterior parts facing in opposite directions (left).

The photographs of partial hydranths are enlarged to 4X the diameter of the whole hydranths.

interact, or they interact very slowly. Each goes its own way, little affected by the other.

A distal part, D, of a primordium in the reactive stage was grafted to the distal end of a whole primordium, DP. The combination is represented, DDP. At the time of combination, these were so close to the stage of tentacle formation that if isolated rather than combined both would have continued to form tentacles. The D would have formed the distal ring of tentacles, and DP would have formed a whole hydranth with both rings of tentacles. When combined as DDP, if everything continued to develop there would have arisen two rows of distal tentacles and one row of proximal tentacles. This never happened. The D at the distal or anterior end (Fig. 7-5) did form distal tentacles in a few hours. The second D region and the P behind it did not continue along their original pathways. Cells beginning to concentrate in bands which would become tentacles dissociated. Several hours later they reunited but in a new pattern. They were following the rule that they could make only those structures not forming anterior to them. They climbed the ladder distally—or anteriorly in this case—until no more rungs were free. This combination of DDP invariably yielded a perfect hydranth with all levels represented but no level represented more than once.

The result was quite different if the D graft was turned around before

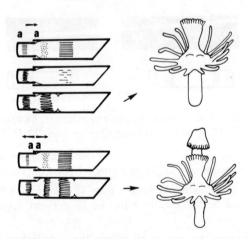

FIG. 7-5. THE DIFFERENT RESULTS WHEN DISTAL GRAFTS AND WHOLE HOST PRIMORDIA ARE JOINED WITH THE SAME AND WITH OPPOSED POLARITIES [*from S. M. Rose*, Jour. Gerontology 22: 30 II, 1967; *full description in text*].

being grafted to the DP host. The combination is represented ⅅDP (Fig. 7-5). Even though firmly attached, neither graft nor host controlled the other. Two anterior ends formed mouth to mouth and the host on the right became a complete hydranth. It turns out from many grafting experiments that a distal end can control in only one direction. If two are combined, ⅅD, controlling information seems to flow away from the level of junction and development is completely independent. This may be the key to the nature of the communication system. We shall return to it.

There is clearly regional specificity to this control. The more levels or parts of levels represented in the graft, the fewer in the host if graft and host are properly aligned. The second P of DPP is always suppressed. D completely prevents the formation of another D downstream and P does the same for P (Fig. 7-6).

An isolated P—that is, a P without a D forming in front of it—itself becomes a D. In the combination DP the P to D transformation is suppressed

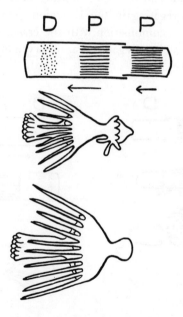

D P P

FIG. 7-6. PROXIMAL SUPPRESSING PROXIMAL [*from* S. M. Rose, Jour. Morph. *100*: 187, 1957].

In the DPP combination, even though the second P was quite far along the pathway of differentiation and continued to develop for a short time after being grafted, it eventually lost all structure.

by the D already present. We learn from the study of these combinations that a whole of many parts arises as each posterior level is reduced to producing the most anterior part not being produced anterior to it.

Grafts do something in addition to suppressing like development behind them. They can also change the polarity of an adjacent portion of the host. In a DPD combination, the second D may remain D or it may lose its organization and reorient as Ɑ with the new tentacles facing the posterior opening. Whether it remains D or becomes Ɑ, it can change the P region so that it becomes Pꟼ. The two results obtained were DPꟼD and and DPꟼƊ (Fig 7-7).

In the DPD combination, what one saw whether the D graft was repolarized by the opening or not was that part of the P region formed new structures oriented with respect to the graft. A row of proximal tentacles faced the host D to the left and another row faced the D or Ɑ to the right. There had been repolarization of that part of P adjacent to D or Ɑ.

Realignment of part of a host also happens sometimes when a ꟼ is grafted to the anterior end of a DP. Starting with ꟼDP, one first notices that cells are disaggregating in the D region. This disaggregation may spread throughout the combination with the erasure of all morphological pattern. Several times disaggregation was observed to stop before spreading

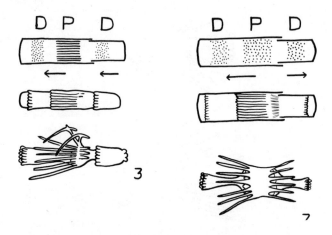

FIG. 7-7. A DISTAL GRAFT CHANGING THE POLARITY OF A PART OF ITS HOST
[*from* S. M. Rose, Jour. Morph. 100: 187, 1957].

Whether a D graft is oriented with the same polarity (left) or with opposed polarity (right) to its host, it can cause part of the proximal region of the host to develop facing the graft.

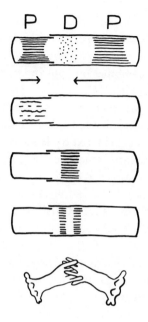

FIG. 7-8. A PROXIMAL GRAFT CHANGING THE POLARITY OF PART OF THE DISTAL
PART OF THE HOST WITH CONSEQUENT CHANGES ALONG THE TWO AXES.
[*from S. M. Rose*, Jour. Morph. 100: 187, 1957)].

The P graft on the left first caused the host to lose visible organization. Eventually,
under the action of the graft, a part of the D region was reversed in polarity. This
yielded two adjacent centers ꓒ and D. Each became normal in size, and as each
region lying behind produced to scale the most distal parts not developing distal
to them, there was nothing left to produce the more proximal parts. Although the
original combination contained two proximal tentacle-forming regions, after the
reorganization both axes ended with gonophores which normally lie between the
distal and proximal tentacles.

Imagine three equally-sized units to begin with, ꟼDP. When this becomes ꓒD
with both Ds normal in size, there was only half a P left for each system. Half a
P could make only its more anterior part back to the gonophore region in front
of the proximal tentacles.

from the D region (Fig. 7-8). The D reorganized, but with two D regions,
one adjacent to and properly aligned with the ꟼ of the graft and the other
aligned with P of the host. In these cases the original D region reorganized
as ꓒD. After that each level in the two differentiating systems became the
most anterior structure not differentiating anterior to it. Each of the two new
distal regions was as large as the original single distal region. The other

parts forming behind the D regions were correspondingly large. There was not enough material to form two complete hydranths on the scale of the anterior parts. The most proximal parts of the two hydranths were missing, even though the original combination had contained two complete partially differentiated proximal regions.

In these last cases D was repolarized by the action of a P graft. In the former cases part of P was repolarized by a D graft. This kind of alignment during differentiation was considered by Ross Harrison to be the result of asymmetrical molecules being oriented (see discussion by Jane Oppenheimer 1966). What could the orienting force be? First of all, we know what it could not be. It could not be a case of one set of aligned molecules causing others to align with respect to them, as in crystallization. This is ruled out by the fact that either D or Œ as noted above caused the same change in polarity in P.

It is not as though each aligned particle caused the alignment of the next. Instead, there is an overall interaction causing realignment at a distance. What could the orienting force be? Could it be electrical? We know that there are bioelectric fields in all organisms.

Reversal of the direction of organization during regeneration in planaria under the action of an applied electric field has already been discussed in Chapter Six. Actually, this experiment was first performed with hydroids. E. J. Lund (1921, 1923) had clearly demonstrated that the hydroid *Obelia* will form hydranths at the end of a stem facing the positive pole of an applied field. Lester Barth (1934) confirmed this with other hydroids, but some species tended to form hydranths in the direction of the negative pole.

Knowledge that non-injurious applied electrical fields in which regeneration was normal could determine the site and direction of transformation from stem to hydranth led Lund to suggest that the normal bioelectric fields may in some way control morphogenesis. Although Lund and others had demonstrated a relationship between electrical fields and morphogenesis, this could not be integrated with other knowledge concerning the origin of patterned organization. Because knowledge, especially at the molecular level, was so limited, there was a tendency to suppose in a fuzzy way that bioelectricity was simply a product of life and that it played no part in determining organization.

Let us consider just what the electrical studies demonstrated. Without an applied current, distal parts of isolated stems of *Tubularia* about 5–10 mm long transform to hydranths and the hydranths almost always face the distal opening. When we place stems shortly after isolation in a weak field of between 50 and 100 microamperes per square millimeter ($50–100\mu a/mm^2$)

with a voltage drop of 10 volts in 10 centimeters (1v/cm), they show normal regeneration and always at the distal end when facing the negative pole. When the field is reversed, 50 percent or more of the stems develop with reversed polarity. The hydranths form at the proximal end facing the opening and no hydranths form in the original distal position. Differentiation and dominance have changed in direction with the result that the hydranth faces the negative pole.

A new hypothesis took the form of this question: "Is the direction of differentiation and dominance reversed because the electric field moves the regionally specific repressors in the opposite to normal direction?" This hypothesis is incorrect as we shall see in a moment, but the experiments which it generated have given us clearer understanding.

The first experiment was a modification of the DDP experiment. Without an applied field, the D on the left would form D structures. The second D and the P would then have to become those structures not forming ahead of them. When DDP faced the positive pole, that is, (+)DDP(−), the same thing happened. When the arrangement was (−)DDP(+), both Ds formed D structures (Fig. 7-9). The anterior D had not suppressed the more posterior D. A two-headed hydranth was produced. At first sight this result might seem to conform with the hypothesis, but it does not. If it did, the more posterior D should have repressed the more anterior D. If the distribution of repressors were simply by electrophoresis. the D closer to one pole should always repress the other. The fact that the applied field blocked the control in one direction but did not make it move in the other direction requires further consideration.

Something else which requires an explanation is that when an applied field changes the polarity of *Tubularia*, the repolarized system faces the negative pole. This causes the surface facing the negative pole to become the most negative locus. Internally the sign is reversed and the most positive internal locus faces the negative pole (Rose, 1970b). Repressive control moves from the more positive internal regions toward more negative regions.

It was already known from Lund's and Barth's experiments on hydroids and from the work of Marsh and Beams with planaria that polarity can be changed in an electrical field. Why did this not happen in the DDP experiments? In the earlier work in which polarity had been changed, the parts regenerated in a field from the beginning and for a long time. Under those conditions, the alignment could be determined by the field. In the DDP combinations both D and DP had been allowed to develop outside the field up to the time when they were combined. Their polarities had already been

F<small>IG</small>. 7-9. B<small>LOCKING SUPPRESSION OF A DISTAL REGION LYING BEHIND ANOTHER BY</small>
AN <small>ELECTRICAL FIELD</small> [*from* S. M. *Rose*, Jour. Gerontology, 22: II, 28,
1917].

In these cases, the starting materials were a distal graft on a whole primordium
containing both D and P, that is, DDP. When the arrangement was $+$DDP$-$
(*right*), the second D was suppressed and became P and the original P region of
the host became stem. The result was a normal hydranth.

The same combination but with the sign of the field reversed, $-$DDP$+$, resulted
in two Ds and one P.

The white lines in the background of the photograph on the left separate the two
self-differentiated D regions.

established during the early stages of regeneration. The polar differences
were visible when the experiment began. Primordia were beginning to take
form at the distal but not at the proximal ends. By design, the combinations
were in the field only for the few hours when the effective regional interaction is known to occur. This is just before and during the early visible
stages of tentacle formation.

There are two facts to be considered. First, the control of D1 over D2
was blocked when the joined pair faced the ($-$) pole. The passage of some
kind of controlling information was blocked in an already polarized system.
The other fact is that control was not forced in the wrong direction by the
field ("wrong" here means against the already established polarity).

It would appear that early in regeneration polarity can be determined
by an applied field and possibly by a bioelectric field. The nature of the
structural polarity is not known. It might be something like a synapse or

valve with transport possible in only one direction. Whatever it is, messages can be carried in one direction. This experiment tells us nothing about the nature of the force moving the messages.

The fact that these messages could be blocked in passing from D1→D2 when the situation was (−)D1D2(+) indicated the possibility that the message was borne by a particle or molecule bearing a positive charge. This led to the suggestion that the regional repressors might be separated by electrophoresis.

Before getting into the matter of separation of repressors, the effect of supernatants of homogenates after centrifugation at 5000xg should be considered (Rose and Powers, 1966). The starting materials were distal halves and proximal halves of either adult hydranths or partially regenerated hydranth primordia. The effect of these regional supernatants was tested on advanced regenerates whose distal parts, hypostome and distal tentacles, had been removed. These pieces were chosen because Hans Driesch (1897) had shown that part of such a recently regenerated proximal part of a hydranth could often regenerate the missing distal region in a few hours, sometimes in as little as 4 hours. When these already differentiated proximal parts were cultured with effective concentrations of materials from fully differentiated proximal regions, their organization was lost and they became a mass of cells. Proximal substances were able to cause already developed proximal regions to lose their structure. Distal substances did not cause proximal regions to lose structure, but they did prevent distal regeneration. Proximal regions simply healed over and failed to produce distal structures for many hours.

The effect of these regional substances does not depend upon a quantitative difference. There was no dilution of either distal or proximal extract which mimicked the effect of any dilution of the extract of the other region.

In the above experiment, the test pieces were fully formed young hydranths each already showing a deep constriction between the hydranth and the stem from which it had arisen. The result was different when younger test pieces were used. These younger pieces were from primordia which had not yet reached the constricted stage. Some were isolated in sea water, others in proximal supernatants and still others in distal supernatants. Most of these younger proximal pieces, whether isolated in sea water or in sea water with proximal materials, lost all structure. The result was different when the young proximal pieces were cultured with distal materials. These test pieces did maintain their proximal structures and continued to develop but without forming new distal structures until a number of hours later.

Isolated proximal regions are always ready to transform to distal regions. In younger primordia this involves a complete loss of visible organization. In older pieces most of the P structure is maintained and only their most distal part loses structure in the process of transforming to D. This is a situation in which proximal structure develops and is maintained only in the presence of distal materials. In the absence of distal materials a P isolate may transform to D if small and to DP if larger.

It is quite clear that all levels can and will transform to distal structures if not prevented from doing so by receiving distal materials. In former times it might have been said that D induces P. By that it was meant that positive information—an "inductor"—was produced by D, which caused other cells to make P structures. All we really know is that an already established D in the proper position and alignment with respect to a differentiating region can prevent that new region from becoming D. The differentiating region has within the genome of its own cells information for producing everything. There is no evidence that D tells any other region how to be P. The simplest statement of the findings with extracts is that each region will produce the most distal structures it is not prevented from producing.

We return now to a consideration of a possible relationship between applied electric fields and polarized control. The evidence presented so far indicates that the polarity of regenerating hydroids and planaria can be determined by an electric field. In addition, the control of one region over a potentially like region was blocked if DDP faced the negative pole in a field of 100 $\mu a/mm^2$ with a voltage drop of 1 to 2.5 v/cm. This indicated the possibility of a controlling agent bearing a positive charge. We do not know whether or not such agents are moved in the organism because of that charge. Electric fields can be observed to change polarity, and the direction of control is perfectly correlated with the polarity, Whether electrical fields also move charged substances from cell to cell is not known. Most people who have studied the problem believe that the normal bioelectric fields are not great enough to influence the direction of movement through cell surfaces. However, the amount of energy required might be slight when cells form tight junctions with no apparent extracellular space between them as they do during differentiation (Trelstad, Hay and Revel, 1967).

Electrically charged repressors were sought by the use of starch gel electrophoresis. Homogenates of D, P, and S (meaning stem) were prepared from differentiating primordia of *Tubularia*. These were placed separately in central wells in slabs of starch gel. Electric currents of the dimensions used to move proteins and other substances from the wells were used. After a run of several hours, samples were taken from various positions to the right and left of the wells (Fig. 7-10). To the right were positively charged materials moving toward the negative pole. The negatively charged materials moved to the left toward the positive pole. The samples of starch containing charged substances were taken in the form of cores which could

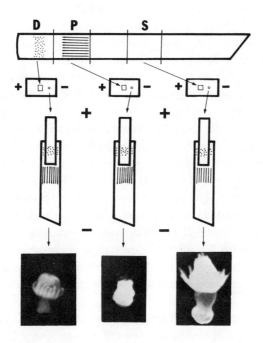

FIG. 7-10. ELECTROPHORETIC SEPARATION AND DELIVERY OF REGIONAL MORPHO-GENETIC REPRESSORS TO PRIMORDIA OF TUBULARIA [*from S. M. Rose*, Jour. Geron·ology, 22: II. 33, 1967].

Stems with hydranth primordia in the stage shown at the top were separated into distal (D), proximal (P) and a stem fraction (S). These were homogenized separately, placed in the central well of a starch slab, and electrophoresed. A small core was taken from the negative region to which positively charged substances had migrated. The cores—(now shown at higher magnification)—were inserted in stems with primordia in a weak electric field.

Below are photographs of typical results. On the right is a complete hydranth unaffected by treatment with S^+ materials. In the center is *Tubularia* tissue treated with P^+ materials and unable to transform to hydranth. On the left is a hydranth treated with D^+ materials. It lacks the distal region.

The rule is that D^+ stops D development whereas P^+ stops the homologous P region plus everything anterior to it.

be inserted into ends of regenerating stems of *Tubularia*. The starch plus charged substances had no effect on regeneration if simply inserted in the regenerating stems. It was found necessary to move the substances out of the starch and make them available to the *Tubularia* by applying weak, non-injurious electric fields.

When the separations were performed at pH 8.6, positively charged materials were discovered which had the effects seen in the photographs of Fig. 7-10. Positively charged S materials had no effect or caused slight delay of hydranth development. P materials blocked all further differentiation of both P and D regions and caused regression. D materials blocked D structure formation while allowing P to continue its development. In some cases P structures formed at the anterior end of the regenerate under the influence of D substances just as if a D graft had been in front. The positively charged substances duplicated the effect of a graft.

The active agents could sometimes also be found to the left of the wells (Fig. 7-10) when electrophoresis had occurred at pH 8.6–8.9 (Rose, 1966, 1967a and b; Gwynn and John Akin, 1967). This indicated that at these pHs the substances might be near their isoelectric points and have almost balanced positive and negative charges. Even water eddies around the wells could have moved some of the effectively neutral particles.

The active agents appear to be proteins or polypeptides or to be carried by such molecules because all inhibitory activity disappeared when small amounts of trypsin were added before electrophoresis (Rose, 1966; Akin, and Akin 1967).

In the pH range 8.6–8.9 all stainable proteins moved toward the positive pole whereas the inhibition moved toward the negative pole. This indicated that, if the repressors were proteins, they were in very low concentration and that they are very basic.

At pH 4.1 the stainable proteins as well as the inhibition moved toward the negative pole. The main difference at pH 4.1 was that all regional specificity disappeared. Stem materials were as inhibitory as those from D and P, and all blocked transformation from stem to hydranth completely. This may be an important clue.

The best evidence that the regional repressors are at least in part histones has been provided by Pierre Lenicque and Marianne Lundblad (1966a, b). They learned that there are present in anterior regions of *Hydra* and the hydroid, *Clava*, electropositive materials which depress the regeneration of anterior structures. In *Clava* they have gone further and shown by separation on ion-exchange columns and by gel filtration that the inhibition is found in the fractions containing histones. They have also separated out substances which promote regeneration; more about these later.

Taken together, the various facts including those in Chapter One lead one to believe that histone-RNAs may be the repressive regional control agents. The work leading to the notion that a stretch of DNA is repressed when the complimentary RNA-histone is bound to it was discussed in Chapter One. The regionally limited effects of repressors obtained at pH 8.6 has been observed. The general nature of repression after extraction at pH 4.1 is to be expected if repressors are histones. Under more acid conditions all histone is separated from DNA. One would expect by this method to obtain from the chromosomes of any region the total variety of histone-RNAs which would block the action of much or all of the DNA thus interfering with regeneration of all regions.

The picture which seems to be emerging is that DNA regions, for example the regions involved in making D structures, release repressors which move in the morphogenetic field and repress like stretches of DNA in cells downstream. Another region, the S region, would not be releasing D repressors because it is inactive as the result of those regions already being repressed. If, however, the S region were extracted with acid, D and other repressors would be released from the chromosomes. Under natural conditions specificity of control would depend upon the release of a few repressors.

The above work with electrophoretically separated repressors started with 60 or more pieces of primordia per well. How much material from how many primordia was necessary is not known. If differentiation really proceeds as in the way suggested, one D region would have to produce enough D repressor to control a potential D region behind it. The knowledge that D repressors can be moved by an applied electric current was utilized. The D + DP experiment was repeated, but this time D and DP were not allowed to touch and their cells to fuse. Instead, sea-water–agar cylinders were inserted in graft and host, and the two were brought close together on the cylinder passing through them, but they were not allowed to touch (Rose, 1967a) (Fig. 7-11). If such preparations were simply left in sea-water, no effect of graft on host was noted. Both graft and host produced anterior structures. When the preparations faced the negative pole in fields of 22 $\mu a/mm^2$ and approximately 1 v/cm, there was little or no effect of graft on host. There was considerable effect when the preparation faced the positive pole. A D graft could affect the DP host across a 0.1 mm agar–sea water gap.

When DP hosts were relatively old primordia, the D graft could block

D in the host while allowing P to proceed (Fig. 7-11). When DP hosts were younger, total dedifferentiation usually resulted from the treatment. Sometimes younger whole DP primordia escaped total dedifferentiation and reacted to the graft across the gap just as if the two had been firmly attached. D of the graft developed, but the D region of the host became P and P

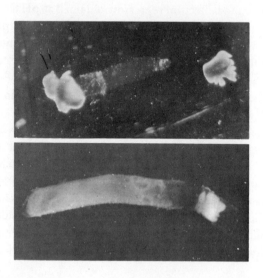

FIG. 7-11.   THE EFFECT OF GRAFTS ACROSS SEA-WATER-AGAR BRIDGES WHEN FACING THE POSITIVE POLE [*from S. M. Rose*, Growth 31: 149, 1967a].

Above: Graft and host had grown close together on an agar skewer facing the positive pole to the right. For the photograph, they had been removed from their perisarcs and their skewer. The graft on the right is the distal region with the ring of distal tentacles and some tissue behind them. The host on the left had been a complete D+P primordium, but produced only P structures including proximal tentacles, base of hydranth, and stem. The host had produced all parts which had not developed in the graft.

Below: Graft and host had been left in their perisarcs and in place on their agar skewer (almost invisible). The graft on the right was a proximal part of a primordium and a short stretch of stem. It had developed into the base of a hydranth on a short piece of stem, and its distal part at the far right had formed a bud which was in the process of transforming to the missing distal structures.

The host had produced neither distal nor proximal structures. There had been appreciable cellular disaggregation in the region of the primordium, leaving a Swiss-cheese effect.

This and similar experiments have demonstrated that the regional repressors are positively charged at the pH of sea water—approximately pH 8.1

became S. The result was that the host produced only that part of the hydranth not produced by the graft. Together they produced one complete hydranth (Fig. 7-11).

P grafts separated from their DP hosts also provided +substances which were transported to and effected a change in their hosts. The P grafts blocked both P and D (Fig. 7-11). Never was P alone suppressed in the host.

For the sake of simplicity, the primordium has been treated as having only two parts, D and P. Actually a hydranth varies in form and structure throughout its length. The regions under control of different parts of the genome are not as numerous as in a mammal, but there are certainly more than two. The grafting work in general, whether there be gaps or not, indicates that the host can produce all of the more proximal levels not produced by the distal graft.

In the gap experiments the specific regional information is supplied by the graft and is transported across the gap by an electric current. As discussed above, there is no evidence that the unidirectional transfer of control from cell to cell is by an electric current. However, we shall see again that the polarity of the transport system can be determined by an electric current.

Understanding of how this change in polarity may affect morphogenesis is based upon the fact that an isolated young P region is unstable and will not continue to develop as a P. Such P pieces lose all organization and often transform to D structures. However, if they are grafted to D pieces or receive D materials by electrical transport from D pieces or even if they are simply bathed in rather large concentrations of D homogenates, the P structure may be retained and P development proceed. P structure is attained and maintained because D suppressors are present. If they are not, P loses P structure as a first step in transformation to D. This process of dedifferentiation as a first step in transformation to something more distal seems to be a universal phenomenon (earlier chapters).

When primordia are kept continuously from very early stages in electric fields of the dimensions used in these studies, they may complete normal development and emerge as young adult hydranths if they face the positive pole. If they face the negative pole the more proximal regions may fail to develop while the more distal regions continue to develop. In this way one can get hydranths without a base, or without base and proximal tentacles, or with too few proximal tentacles. In some cases there is regression of structures already formed in P regions. This failure of P to develop and its tendency to regress when DP faces the (—) pole is probably a sign of failure of D+ substances to reach the proximal regions in sufficient quantity. It might be that some of the failure was because the applied field was forcing the positive substances in the wrong direction, but it appears more likely that it was due to a realignment of polar molecules. This realignment may be at cell surfaces where transport between cells is effected. This important aspect of the control of differentiation awaits study.

Throughout studies of development, cases of linear control have appeared. Possibly because of the inclusion of other substances, apical auxin applied to one side of the top of a decapitated coleoptile of an oat seedling causes the cells below on the same side to elongate. Even before fibrovascular bundles have differentiated, the material does not spread uniformly in all directions but is transported more along one line than along others. In the development of the amphibian limb, the pathways of control as worked out by Harrison are also linear.

Here, too, in *Tubularia*, the control is linear. This has been apparent in several cases when inserted starch cylinders with included natural inhibitors made contact on only one side. Transformation was suppressed on the side in contact, and it remained stem whereas a half hydranth arose on the side where contact had failed (Fig. 7-12).

Evidence has been presented indicating that imposed electric fields can determine the direction of the polarized transport system. It is believed that bioelectric fields may also determine molecular orientation in developing systems. If so, the case of interference when two regenerating limbs are joined at an angle becomes understandable (Chapter Three). One would expect a jumbled arrangement rather than an orderly molecular pattern along which controlling information could travel. Here again is an area where study should be profitable.

Following the pioneering work of Professor Child and a clear synthesis of his and related work by Julian Huxley and G. R. DeBeer (1934, reprinted 1963), it became apparent that differentiation occurs along quantitative gradients. It was further established both in plants and in animals that as differentiation proceeds in a given region, the same differentiation is precluded in other parts of the same polarized field. This far all of the investigators agree. As always there are differences at the forefront of our knowledge.

There has been a tendency among some of the modern students of regeneration in *Hydra* to think of the level of differentiation as resulting from interaction of inhibition and another factor such as stimulation (Allison Burnett, 1966; Julian Haynes, 1967) or threshold to inhibition (Gerald Webster and Lewis Wolpert, 1966: Webster, 1966a and b). A

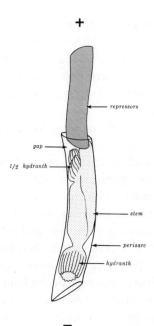

FIG. 7-12.   LINEAR REPRESSION BY P⁺ MATERIALS [*from S. M. Rose*, Growth 31: 149, 1967a].

This is a case from the experiment diagrammed in Fig. 7-10 where the tissue of the primordium made contact with the inserted starch on only one half of the cylinder. P⁺ repressors moving toward the negative pole prevented regeneration only in the region of contact and directly behind it. There was no appreciable lateral movement of repressors, and the part separated from the starch by a gap became half a hydranth. At the proximal end of the stem, farther away, a complete hydranth formed facing the proximal opening.

particular level of differentiation would depend upon a ratio between an inhibitor or inhibition and the other factor.

Stimulation of regeneration is difficult to assess and stimulators have not been isolated. A variety of things like nucleoprotein (Lenicque and Lundblad, 1966a), increased temperature, or dilution of the medium may increase the rate of regeneration but these cannot be considered specific stimulators or initiators. Actually, all one has to do to "stimulate" regeneration in all kinds of regenerating organisms is remove a part or block an axis of control so that a former downstream region becomes the most upstream region remaining. Another way is to set up an opposing axis. Child (1927) did this with the hydroid *Corymorpha* by gashing the side. Here nothing was added or removed, still a new axis of regeneration was established at the point of injury. Also, new axes can be established by the deviation of nerves (Chapters Three and Six). Stimulators or initiators of regeneration appear to be superfluous in any system in which simple isolation or being at the distal end of a system leads to transformation to the most distal part of that system.

The gradients seem to result in polarized axes along which differentiation can occur as controlling information is passed. Confusion has resulted from many studies in which it was not realized that the establishment of axes and differentiation along the axis are different processes. Quantitative gradients seem to have much to do with axes but may have nothing to do with repression which occurs along an axis. Apparently it is not a ratio which represses a particular gene but a qualitatively specific repressor. These repressors may move from cell to cell because of gradients and associated polarized transport, but there is no knowledge to suggest that a particular part of a gradient causes a particular kind of differentiation.

Since this chapter was written, new evidence has appeared, establishing that a bioelectric field is necessary for development. Stems of *Tubularia* were ligated at both ends, and the potential differences along the stems were measured and compared with those in isolated, unligated stems. The ends of the open stems became electronegative to middle regions and remained so until a new hydranth had begun to form. The same thing began to happen in the closed stems, but before regeneration could occur the ends had lost their relatively high negativity (-0.9 to -1.7 millivolts) and were less negative or even became positive. Some of the ligated stems were then placed in electric fields where there was a difference in potential of approximately 1 volt per centimeter. Some of the stems regenerated even under their ligatures and some did not. Those which did regenerate had been given potential differences as great as they normally produce when not ligated. In some cases the artificial field reversed the original electrical polarity of the stems. In these cases morphogenesis was also reversed. Hydranths arose from ends which had originally been the posterior ends (Rose, 1970a,b).

These results indicate that a certain level of potential difference is necessary if the cells are to communicate and take part in regeneration. If their own natural field is diminished, regeneration fails. It may be reinstated by an applied electrical field.

It has also been demonstrated in the alga, *Fucus*, that a certain level of potential difference is required for differentiation. Friedrich-Wilhelm Bentrup (1968) first determined the intensity of blue light required to polarize the egg and cause a rhizoid to appear. He then determined the drop in voltage which would do the same thing. Next, the eggs were subjected to the same amounts of the two polarizing agents but in opposite directions. The two working against each other canceled each other and rhizoids did not form. The eggs remained spherical. This result considered with those discussed in Chapter Four indicates that the various polarizing agents all lead to the establishment of electrical differences sufficient to move messages from cell to cell.

From the work on *Tubularia* it is now clear that the bioelectric field is the polarized message transport system. The messages are produced by the cells, and as the result of their passage down the one-way road, the cells down the road cannot produce those structures forming up the road. They are free to use their most up-road genes not being repressed.

References Cited

Akin, G. C. 1966. Self-inhibition of growth in *Rana pipiens* tadpoles. *Physiol. Zool.* 39: 341–356.

Akin, G. C. and Akin, J. 1967. The effect of trypsin on regeneration inhibitors in *Tubularia*. *Biol. Bull.* Woods Hole 133: 82–89.

Baker, J. R. 1952. *Abraham Trembley of Geneva.* 259 pp. Edward Arnold & Co.

Barth, L. G. 1934. The effect of constant electric current on the regeneration of certain hydroids. *Physiol. Zool.* 7: 340–364.

———. 1938a. Quantitative studies of the factors governing the rate of regeneration. *Biol. Bull.* Woods Hole 74: 155–177.

———. 1938b. Oxygen as a controlling factor in the regeneration of *Tubularia*. *Physiol. Zool.* 11: 179–186.

———. 1940. The process of regeneration in hydroids. *Biol. Rev.* 15: 405–420.

Belousov, L. B. and Geleg, S. 1960. Chemical regulation of the morphogenesis of hydroid polyps. *Doklady Akad. Nauk SSSR* 130: 1165–1168.

Bentrup, F. W. 1968. Die Morphogenese pflanzlicher Zellen im elektrischen Feld. *Z. Pflanzenphysiol.* 59: 309–339.

Braem, F. 1908. Die Knospung der Margeliden, ein Bindeglied zwischen geschlechtlicher und ungeschlechtlicher Fortpflanzung. *Biol. Zbl.* 28: 790–798.

Brien, P. and Renier-Decoen, M. 1949. La croissance, la blastogénèse, l'orogénèse chez *Hydra fusca* (Pallas) *Bull. Biol. France Belg.* 83: 293–386.

————. 1955. La signification des cellules interstitielles des Hydres d'eau douce et la problème de la réserve embryonnaire. *Bull. Biol. France Belg.* 89:258–325.

Browne, E. 1909. The production of new hydranths in *Hydra* by the insertion of small grafts. *J. Exp. Zool.* 7:1–24.

Burnett, A. L. 1966. A model of growth and cell differentiation in Hydra. *Amer. Nat.* 100:165–190.

Campbell, R. D. 1967a. Growth pattern of *Hydra*: distribution of mitotic cells in *H. pseudoligactis*. *Trans. Amer. Microsc. Soc.* 86(2): 169–173.

————. 1967b. Tissue dynamics of steady state growth in *Hydra littoralis*. 1. Patterns of cell division. *Devel. Biol.* 15:487–502.

Child, C. M. 1927. Experimental localization of new axes in *Corymorpha* without obliteration of the original polarity. *Biol. Bull.* Woods Hole 53:469–480.

————. 1929. Physiological dominance and physiological isolation in development and reconstitution. *Arch. Entw-mech. Org.* 117:21–66.

————. 1941. *Patterns and Problems of Development*. 811 pp. Chicago, Illinois: Univ. of Chicago Press.

Chun, C. 1896. Atlantis Biologische Studien über pelagische Organismen. *Biblio. Zool.* H. 19.

Diehl, F. A. and Burnett, A. L. 1965. The role of interstitial cells in the maintenance of *Hydra* III. Regeneration of hypostome and tentacles. *J. Exptl. Zool.* 158:299–318.

Driesch, H. 1897. Studien über das Regulationsvermögen der Organismen. 1 Von den regulativen Wachstums- und Differenzirungsfähigkeiten der Tubularia. *Arch. Entw-mech. Org.* 5: 389–418.

Fulton, C. 1959. Re-examination of an inhibitor of regeneration in *Tubularia*. *Biol. Bull.* Woods Hole 116:232–238.

Gilchrist, F. G. 1937. Budding and locomotion in the Scyphistomas of *Aurelia*. *Biol. Bull.* Woods Hole 72:99–124.

Godlewski, E. 1904. Zur Kenntnis der Regulationsvorgänge bei *Tubularia mesembrianthemum*. *Arch. Entw-mech. Org.* 18:111–160.

Hay, E. D. 1966. *Regeneration*. Holt, Reinhart and Winston, New York, pp. 148.

Haynes, J. F. 1967. The analysis of species specific variations in form and growth pattern in *Hydra*. *J. Embryol. Exptl. Morph.* 17:11–25.

Haynes, J. and Burnett, A. L. 1963. Dedifferentiation and redifferentiation of cells in *Hydra viridis*. *Science* 142:1481–1483.

Huxley, J. S. and de Beer, G. R. 1934. *The elements of experimental embryology*, *Cambridge*. Reprinted 1963, New York: Hafner.

Kanajew, J. 1930. Zur Frage der Bedeutung der interstitiellen Zellen bei Hydra. *Arch. Entw-mech. Org.* 122:736–759.

Langhans, V. H. 1909. Uber experimentelle Untersuchungen zu Fragen der Fortpflanzung, Variation und Vererbung bei Daphniden. *Verh. der Deutsch. Zool. Ges. Frankfurt a.M.* 19:281–291.

Lenicque, P. M. and Lundblad, M. 1966a. Promoters and inhibitors of development during regeneration of the hypostome and tentacles of *Clava squamata*. *Acta. Zool.* 47:185–195.

———. 1966b. Promoters and inhibitors of development during regeneration of the hypostome and tentacles of *Hydra littoralis*. *Acta. Zool.* 47:277–287.

Lund, E. J. 1921. Experimental control of organic polarity by the electric current. I. Effects of electric current on regenerating internodes of *Obelia commisuralis*. *J. Exptl. Zool.* 34:471–493.

———. 1923. Threshold densities of the electric current for inhibition and orientation of growth in Obelia. *Proc. Soc. Exptl. Biol. Med.* 21:127–128.

Miller, J. A., Jr. 1937. Some effects of oxygen on polarity in *Tubularia crocea*. *Biol. Bull.* Woods Hole 73:369.

———. 1959. Nutritive substances and reconstitution in *Tubularia*. *Proc. Soc. Exptl. Biol. Med.* 100:186–189.

McCallum, W. B. 1905. Regeneration in plants. *Bot. Gaz.* 40:97–120, 241–263.

Moore, J. A. 1939. The role of temperature in hydranth formation in *Tubularia*. *Biol. Bull.* Woods Hole 76:104–107.

Normandin, D. K. 1960. Regeneration of *Hydra* from the endoderm. *Science* 132:678.

Oppenheimer, J. M. 1966. Ross Harrison's contributions to experimental embryology. *Bull. of History of Medicine.* 60:525–543.

Rand, H. W., Bovard, J. F., and Minnich, D. E. 1926. Localization of formative agencies in Hydra. *Proc. Natl. Acad. Sci.* U.S. 12:565–570.

Richards, C. M. 1958. The inhibition of growth in crowded *Rana pipiens* tadpoles. *Physiol. Zool.* 31:138–151.

Rose, S. M. 1957. Polarized inhibitory effects during regeneration in *Tubularia*. *J. Morphol.* 100:187–205.

———. 1960. A feedback mechanism of growth control in tadpoles. *Ecology* 41:188–199.

———. 1966. Polarized inhibitory control of regional differentiation during regeneration in *Tubularia*. II. Separation of active materials by electrophoresis. *Growth* 30:429–447.

———. 1967a. Polarized inhibitory control of regional differentiation during regeneration in *Tubularia*: III. The effects of grafts across sea water-agar bridges in electric fields. *Growth* 31:149–164.

———. 1967b. The aging of the system for the transmission of information controlling differentiation. *Jour. Geront.* 22(II):28–41.

———. 1970a. Restoration of regenerative ability in ligated stems of *Tubularia* in an electrical field. *Biol. Bull.* Woods Hole 138:344–353.

———. 1970b. Differentiation during regeneration caused by migration of repressors in bioelectric fields. *Am. Zoologist* 10:91–100.

Rose, S. M. and Powers, J. A. 1966. Polarized inhibitory control of regional differentiation during regeneration in *Tubularia*. I. The effect of extracts from distal and proximal regions. *Growth* 30:419–427.

Rose, S. M. and Rose, F. C. 1941. The role of a cut surface in *Tubularia* regeneration. *Physiol. Zool.* 14: 328–343.

——. 1961. Growth-controlling exudates of tadpoles. In *Mechanisms in biological competition*, ed. F. L. Milthorpe. *Symp. Soc. Exptl. Biol.* 15: 207–218.

——. 1965. The control of growth and reproduction in freshwater organisms by specific products. *Mitt. Internat. Verein Limnol.* 13: 21–35.

Slautterback, D. B. and Fawcett, D. W. 1959. The development of the cnidoblasts of *Hydra*. An electron microscope study of cell differentiation. *J. Biophys. Biochem. Cytol.* 5: 441–452.

Steinberg, M. S. 1954. Studies on the mechanism of physiological dominance in *Tubularia*. *J. Exptl. Zool.* 127: 1–26.

——. 1955. Cell movement, rate of regeneration and the axial gradient in *Tubularia*. *Biol. Bull.* 108: 219–234.

Steinberg, S. N. 1963. The regeneration of whole polyps from ectodermal fragments of scyphistoma larvae of *Aurelia aurita*. *Biol. Bull.* Woods Hole 124: 337–343.

Stevens, N. M. 1902. Regeneration in *Tubularia mesembrianthemum*. *Arch. Entw-mech. Org.* 13: 410–415.

Tardent, P. 1955. Zum Nachweis eines regenerationschemmenden Stoffes im Hydranth von Tubularia. *Rev. Suisse Zool.* 62: 289–294.

——. 1960. Principles governing the process of regeneration in hydroids. In *Developing cell systems*, ed. D. Rudnick. New York: Ronald Press. p. 21.

——. 1963. Regeneration in the Hydrozoa. *Biol. Rev.* 38: 292–333.

——. 1964. Der Sauerstoff-Verbrauch normaler und regenerierender Hydrocauli von *Tubularia*. *Rev. Suisse Zool.* 71: 167–181.

——. 1965. Developmental aspects of regeneration in coelenterates. In *Regeneration in animals and related problems*, ed. V. Kiortsis and H. A. L. Trampusch pp. 89–94. Amsterdam: North-Holland Publ. Co.

Tardent, P. and Eymann, H. 1959. Experimentelle Untersuchungen über den regenerationshemmenden Faktor von *Tubularia*. *Arch. Entw-mech. Org.* 151: 1–37.

Tokin, B. P. and Gorbunowa, G. P. 1934. Untersuchungen über die Ontogenie der Zellen II. Wie aus einem Stiel von Hydra fusca eine ganz Hydra experimentell regenerieren kann. *Biol. Zhur.* 3: 306.

Trelstad, R. L., Hay, E. and Revel, J. P. 1967. Cell contact during early morphogenesis in the chick embryo. *Devel. Biol.* 16: 78–106.

Trembley, A. 1744. Mémoires pour servir à l'histoire naturelle d'un genre de polypes d'eau douce, à bras en forme de cornes. Leyden.

Tweedell, K. S. 1958a. Inhibitors of regeneration in *Tubularia*. *Biol. Bull.* Woods Hole 114: 255–269.

——. 1958b. A bacteria-free inhibitor of regeneration in *Tubularia Biol. Bull.* Woods Hole 115: 369.

Vernon, H. M. 1899. The relations between marine animal and vegetable life. *Mitt. Zool. Sta.* Neapel 13: 341–425.

von Bertalanffy, L. and Rella, M. 1941. Untersuchungen zur Gesetzlichkeit des Wachstums. VI Teil. Studien zur Reorganization bei Süsswasserhydrozoen. *Arch. Entw-mech. Org.* 141: 99–110.

Webster, G. 1966a. Studies on pattern regulation in hydra. II. Factors controlling hypostome formation. *J. Embryol. Exptl. Morph.* 16: 105–122.

———. 1966b. Studies on pattern regulation in hydra. III. Dynamic aspects of factors controlling hypostome formation. *J. Embryol. Exptl. Morph.* 16: 123–141.

Webster, G. and Wolpert, L. 1966. Studies on pattern regulation in hydra I. Regional differences in time required for hypostome determination. *J. Embryol. Exptl. Morph.* 16: 91–104.

Weismann, A. 1892. *The germ plasm.* English translation by W. N. Parker and H. Rönnfeldt. New York: Scribners.

Zwilling, E. 1939. The effect of the removal of perisarc on regeneration in *Tubularia crocea. Biol. Bull.* Woods Hole 76: 90–103.

———. 1963. Formation of endoderm from ectoderm in Cordylophora. *Biol. Bull.* Woods Hole 124: 368–378.

CHAPTER EIGHT

The Origin and Operation of Polarized Control in Embryos

So far we have been dealing with the controls which operate during the replacement of missing parts in adult organisms. Now we turn to an examination of the origin and operation of these controls during early development in eggs and embryos. A little of the history of experimental embryology is necessary as a background for the understanding of the modern hypotheses and problems.

First we have to know what we are starting with in the egg cell and in the sperm. This part of the study has a fascinating history. In the early days of the microscope when the optics were poor and the imaginations good, the investigators polarized in two groups, the ovists and the spermists. Both believed in preformation, with the ovists thinking of the egg as containing a miniature adult. One of these gentlemen, Gautier d'Agoty, a spermist, thought he could distinguish between sperm of horse and of donkey by the length of the included ears. The beliefs of the preformationists now seem quaint, but one should realize that some very able people held these beliefs and that they had some evidence for them. One observation, a true one, did much to convince the investigators of the 18th century of the validity of the belief. Charles Bonnet (1769) knew that when some Daphnia first emerge from their mother's brood pouch, they, the second generation, already have the third generation in their pouches. Three generations were known to exist in one body. It was from this observation that the theory of emboîtement or encasement arose. All life was formed in the beginning. All the generations were packed within the original parents, box within box.

Belief in preformation in its early gross form gave way to a more subtle form as stages in development were studied. Caspar Friedrich Wolff (1759) usually receives the credit for correcting the early mistake. He reasoned that the intestine of a chick could not possibly be preformed because he could follow its development from a sheet of tissue which folded on itself and became a tube (see Joseph Needham, 1934, for the early history of embryology).

A more subtle form of preformation had its peak expression in the last part of the nineteenth and the early part of this century. It was believed that different kinds of nuclei or cytoplasm are segregated and determine the fates of the parts containing them. August Weismann's theory of differentiation by nuclear segregation was considered with reference to the reserve cell theory (Chapter Seven).

Cleaving eggs of frog (Oscar Hertwig, 1893) and of sea-urchin (Hans Driesch, 1893) were compressed between pieces of glass for short periods of time. Spindles were forced to line up parallel to the glass instead of at right angles to it, and the position of resulting daughter nuclei with respect to each other were abnormal. Although quite abnormal patterns of nuclear distribution were caused, development was quite normal. This was the earliest demonstration that differentiation did not result from segregation of nuclear materials.

The next type of evidence against the segregation theory was the finding that single cells of the four-cell stage of the sea urchin egg could develop into a whole embryo (Driesch, 1900). We shall be developing the idea in this chapter that it is possible for each of the first four cells to develop into a whole—although small—embryo, because each of the first four cells has exactly what the whole egg has: a complete set of genes, a transcription and translation system, and a complete North-South gradient system.

One of the clearest demonstrations of equipotentiality was performed by Professor Spemann (1928). He started with fertilized egg cells of a salamander and tied ligatures around them before first cleavage. The hair which he used as a ligature extended like a meridian around the egg, passing through or close to the north and south pole of the single cell. North and south are used rather than the conventional terms, animal and vegetal, because the egg is easier to visualize if one can think of it as something like the earth. The ligature was then tightened until it almost cut the egg in two. The one nucleus was on one side or the other of the constriction. The ligature remained in place until the one nucleus had divided several times. After four cleavages there were 16 cells on one side of the ligature. On the other side was half the cytoplasm of the egg but no nucleus. When the ligature was relaxed, the cell adjacent to the ligature suddenly had a great increase in cytoplasm and its nucleus was free to move over into the half of the egg without a nucleus. The ligature was retightened. This left one hemisphere, the eastern or western, with one half of the original cytoplasm and one nucleus of the 16-cell stage. The other hemisphere contained half of the cytoplasm and 15 of the first 16 nuclei. Had there been segregation of region-determining nuclear substances, one hemisphere should have been able to form something like 15/16 of an embryo and the other at most something like 1/16. Had there been cytoplasmic segregation each half should have formed at the most half an embryo. What actually happened was that both formed complete embryos, proving that differentiation depends on segregation of neither nuclear nor cytoplasmic materials.

At the time when it was becoming clear that segregation is not involved in the differentiation of echinoderm and amphibian eggs, another kind of evidence obtained from the study of development in annelids and molluscs and their allies led to the belief that their eggs are fundamentally different and that cytoplasmic segregation does play a part in their differentiation.

In many annelid and molluscan eggs the first two cleavage cells differ in size. Just before first cleavage, a large south polar area bulges out as the south polar lobe, or simply polar lobe. As the first cleavage occurs, this polar lobe flows into one of the cells, making it considerably larger than the other cell (Fig. 8-1). Later, just before the second cleavage, a second polar lobe bulges out from the larger cell, the one which had received the first polar lobe. This second polar lobe is smaller than the first. It flows during second cleavage into one of the first four cells.

Edmund Beecher Wilson (1904), the famous cytologist and embryologist of the early twentieth century, found this situation experimentally irresistible. He cut away first polar lobes from some eggs of *Dentalium*, a mollusc, and from others he removed the second polar lobe. The development of his "lobeless" embryos was compared with normal development. The larvae of many of the marine molluscs and annelids are known as trochophores. Three easily-recognized features of trochophores can be seen in Fig. 8-1. There is a top-like shape, rows of long cilia around the equator, and a small group of apical cilia at the anterior end which had been the north pole of the egg.

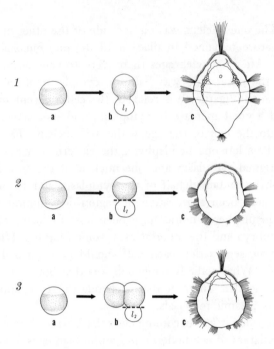

FIG. 8-1. THE EXPERIMENT WHICH LED TO A BELIEF IN DIFFERENTIATION BY
CYTOPLASMIC SEGREGATION [*adapted from E. B. Wilson (1904)*].

1. The egg of the mollusc *Dentalium* (*a*) develops a polar lobe before first cleavage (*b*) and goes on after many cleavages to form a trochophore larva (*c*).

2. When the polar lobe is removed from an egg, the larva is missing its apical tuft of cilia and its posterior region.

3. When the smaller second polar lobe, one which forms just before second cleavage, is cut away, an apical tuft may form but the normal form of the posterior region is not achieved.

One possible explanation is that the material for the apical region had left the lower part of the egg after the first lobe had formed, but before the second formed.

The normal trochophore may be compared with the larvae from which either the first or the second polar lobe had been removed earlier (Fig. 8-1). The removal of the first polar lobe led to the development of a spherical larva with its long cilia bunched together and atypically arranged. Such a larva appeared to Wilson to have developed without a posterior region and without its anterior apical cilia. These he contrasted with the embryos from which only the second polar lobe had been removed. These too lacked a well-formed posterior region but some did have the anterior apical cilia. The suggestion was that materials necessary for development of various parts were bunched together in the original polar lobe. Removal of that lobe took away materials needed for the whole posterior region and the far anterior region. By the time the second lobe appeared, the material for the far anterior apical cilia had moved out of the lobe region up close to the north pole. Consequently, when the second lobe was removed the apical cilia material remained.

With this kind of information only, it was possible to believe in segregation of organ-forming substances or, in agreement with modern thinking, substances which could turn on or turn off certain regional genes.

There were other experimental results with mollusc eggs which were incompatible with the idea of differentiation by segregation. These experiments involved use of rather high-speed centrifuges. Donald Costello (1939) and Anthony Clement (1938) spun mollusc eggs with such force that they were pulled apart into two or more pieces. An occasional fragment of cytoplasm plus nucleus developed into a complete embryo. This could occur even when the fragment was only a part of one of the first four cells. Supposedly, segregation had occurred, yet a part spun off from a part could become a whole. This just did not fit with segregation theory.

Another centrifugation experiment, this one by Arturo Peltrera (1940), teaches us something completely new about the control of development in eggs. Peltrera worked with another mollusc, *Aplysia*. He first separated the early cleavage cells and found, as everyone before him had found, that the cells could continue development, forming some of the structures which they would have formed had they remained a part of a developing egg. These isolated cells never formed as much as they would have as a part of a whole system. When Peltrera took such isolates destined to form partial structures and centrifuged them with a force something over 10,000 times gravity, they were greatly stretched but not torn apart Many became capable of total development after the centrifugation. For example, both of the first two cells, the larger and the smaller, could develop like a whole egg after centrifugation. Each became like a whole egg, and when they divided, each divided like a whole egg into a smaller and a larger cell, and each went on to form a whole embryo. The same was true for each of the first four cells. Each could develop like a whole egg. This means that the origin of differences during development does not result from each cell receiving the materials for special parts. Apparently, they develop differently because they possess different structures. These structures can be obliterated by centrifugation, causing each of the early cells to revert to the relatively unstructured condition of the recently fertilized single cell.

The gross regional structuring comes early in eggs. Again we use the work of E. B. Wilson on *Dentalium*. The first polar lobe forms at the south pole and contains possibly one fourth of the material of the egg. When a northern portion of the egg was cut away shortly after fertilization, the south polar lobe formed in proper proportion to the egg of reduced size. If the whole egg had been reduced to one third size, the lobe was 1/4 of the remaining 2/3. Everything developing from such an egg was smaller but properly proportioned. If the same northern region was cut away from other eggs but not until an hour later, shortly before the first cleavage was expected, the polar lobe was much too large for the reduced mass. Instead, it was the right size for the original unreduced mass. It appears that some time between fertilization and first cleavage the structuring of the egg had occurred, at least to the extent of determining the size of the polar lobe and the future parts of the embryo. Although there is nothing like a stomach present in the egg just before its first division, its size and the size of other organs has been determined by something occurring at that time.

Where could the structure—a structure which can be obliterated by centrifugation almost great enough to pull cells apart—reside? The centrifuge again was the tool used in reaching the answer. When egg cells are centrifuged but with forces less than the great stretching forces used by Peltrera, the contents may be layered. The lighter oils migrate to the centripetal pole and the heaviest materials to the centrifugal pole. At least five layers can be seen in many eggs. Cleavages can occur even when the materials are out of place (Chr. P. Raven, 1958) and normal development result. It has been observed many times by many investigators that once the early structure has been determined, all visible components in the fluid egg cell can be jumbled. For example, several optically different plasms can be observed in the egg of the ascidian, *Styela*. When the egg first reaches the water, it appears to be homogeneous. Shortly after fertilization, the various plasms can be seen to segregate. Each takes a certain position. These are constant positions from egg to egg, and one can predict as Edwin G. Conklin (1905a,b) did that the yellow plasm would always be in the muscles and mesenchyme. Because the mesoderm was always yellow, Conklin first thought that the yellow substance was a mesoderm-forming substance. Later (1931), he showed that the yellow material could be displaced by centrifugation from the region where muscle would form into the region of the future brain. The resulting tadpoles were perfectly normal but with yellow brains and colorless muscles. As far as is known, the position of no cytoplasmic substance has an influence on regionality of development (F. R. Lillie, 1906: T. H. Morgan, 1910; Chr. P. Raven, 1958). As F. R. Lillie and others pointed out early in this century, the polarized structure of the egg must reside in the cortex, as this does maintain its structure and does not come apart during mild centrifugation, which is sufficient to rearrange all of the internal materials.

The original study by E. B. Wilson (1904) which had done so much to convince his contemporaries that differentiation results from cytoplasmic segregation now has to be reinterpreted. It should be recalled that it seemed that some material used in apical tuft formation left the southern cytoplasm between the time of appearance of the first and second polar lobes. The evidence was that removal of the first polar lobe resulted in the absence of the apical tuft, whereas removal of the second lobe did not. Now it is quite certain that this interpretation is not correct. Anthony Clement (1968) has demonstrated with another molluscan egg that primary organization evolves in the cortex rather than in the internal materials.

Clement held eggs upside down in a gelatine–sea water gel during centrifugation. The position of the cytoplasmic ingredients was reversed with respect to the cortex. After upside-down centrifugation, the southern cortex which normally coats a polar lobe when it appears later, contained northern cytoplasm. The eggs were rapidly recentrifuged, but in a medium which allowed the eggs to be pulled apart into two halves. Those halves which contained the southern cortex and northern cytoplasm formed polar lobes and later produced lobe-dependent organs. Those which contained the proper cytoplasm but the wrong cortex did not do so. Clearly early differentiation proceeds in the cortex.

What had appeared to be a case of cytoplasmic segregation in Wilson's work appears to have been a case of progressive cortical differentiation. By the time the second lobe was removed, the organization had progressed sufficiently in the north so that it was no longer dependent upon interaction with the south to produce a future apical tuft site.

Although pattern "sets" in the cortex of eggs between fertilization and first cleavage, the egg in most species is already polarized by the time it leaves the ovary. For example, one pole may have a different color. It is thought that the original polarization may result from differential exposure to oxygen, but this has not been tested rigorously.

The egg of the marine alga, *Fucus*, seems to have no polarity or a very labile polarity when it is deposited in sea water. The direction of the incident light or a CO_2 gradient or other chemical gradients (Douglas

Whitaker, 1931) or the direction of an imposed electric field (E. J. Lund, 1923) can determine the polarity under experimental conditions.

As already noted, patterning, which determines the size of the parts and qualitative regional differences to appear later, occurs before the first division in the eggs of annelids and molluscs. These eggs also exhibit a considerable patterning of visible ingredients at this time (Joseph Spek, 1930, 1934; Erich Ries and Manfred Gersch, 1936). Colored natural inclusions, including some which are pH indicators, are quite homogeneously distributed before fertilization, but segregate with the more alkaline moving to the north pole and the more acid to the south pole. It is quite clear that a system which can move charged particles is being established at this time. Correlated with this system are differences in oxidative metabolism which also exhibit a polarized gradient structure. The important thing which appears to be happening in these eggs before first cleavage is the formation of a polarized transport system.

It is very doubtful because of the evidence given by the centrifuge that any significant amount of controlling information is passed from one region to another while the egg is still a single cell. Instead, it appears that a system is established before cleavage which will transport the information when it is produced later. The part of the system which is retained during mild centrifugation is cortical, but under normal conditions there is visible polarized transport within the more fluid cytoplasm.

The operation of a polarized field can be seen after fertilization of the egg of the ascidian, *Styela partita*. This was first observed by Edwin G. Conklin (1905a) and since by many generations of students at the Marine Biological Laboratory, Woods Hole, Massachusetts. When the eggs and sperm are shed by the adults, the eggs are a uniform very pale yellow. Within minutes of insemination, the yellow concentrates in a south polar cap. The yellow particles move through the cytoplasm.

The egg cells of *Styela* have a fibrous layer, the chorion, extending as the atmosphere does around the earth. There is another membrane, the vitelline membrane, lying very close to the egg surface. Beneath the chorion, and between the chorion and the vitelline membrane, are some satellite cells. Before insemination, these are scattered everywhere all around the egg. Within minutes of insemination and at the same time that the yellow pigment moves through the cytoplasm to the south pole, the satellite cells move quickly to the same pole. The satellites move on the outside of the egg separated from it by a membrane. The observations clearly indicate the establishment of a field with a center at the south pole. It appears to be an electromagnetic field extending through and beyond the surface of the egg similar to the way in which the earth's field extends beyond the surface.

Additional information has been gained concerning the establishment and operation of the morphogenetic axes in the amphibian egg. By the time eggs of frogs and salamanders are laid, they are visibly polarized. The northern hemisphere is darker in shade, and a line running from the middle of the dark area through the egg to the middle of the lighter area is the major axis of the egg. The future anterior end of the embryo will lie close to the middle of the darker field, and the major axis runs essentially in the antero-posterior direction.

By the time the egg has divided into two cells, the other axes have been determined. This was learned by Wilhelm Roux (1883) from one of the first experiments with eggs and embryos. He injured one of the first two cells of the frog's egg with a needle, thereby preventing it from developing further. The other cell either became a mass of undifferentiated cells or after many cell divisions became approximately half of an embryo. It could be a right, or left, or a dorsal half. It could also be a dorso-right or a dorso-left half, but never did purely ventral halves develop. As we shall see, something which appears on the dorsal side is necessary for development. These results first led, as such results always have, to a suggestion of segregation.

The next result was most confusing at first. O. Schultze (1894) inverted eggs and held them in that position during first cleavage. Some of these developed as though each of the first two cells was a whole egg, and joined twins resulted.

When an egg is inverted during first cleavage, an unusual distribution of yolk granules follows. The yolk granules are heavier than the other ingredients and fall to the bottom. The first cleavage furrow begins at the north pole and slowly cuts through the egg until it reaches the south pole. When the egg is inverted, the yolk as it slowly falls through the cytoplasm meets the cleavage furrow rising through the egg. This barrier divides the yolk into two masses. When an egg cleaves in the normal position, the yolk has already fallen into position as a single mass before the cleavage plane passes through it. Very roughly there are in the inverted egg two spheres and in the normal egg two hemispheres.

Later Andreas Penners and W. Schleip (1928) and Penners (1929) demonstrated by tipping eggs that the dorsal center of organization always appears at the junction of yolk and non-yolk. Apparently, if there are two yolk masses, two centers may develop.

More was learned from experiments with half eggs, which reinforces the theory that shape determines axes. J. F. McClendon (1910) not only damaged one of the first two cells as Roux had done, but sucked it out with a pipette from within the vitelline membrane. The remaining cell, half of an egg, could then become a whole embryo. With this result alone, one could think of the dead cell while in place as exerting a chemical or contact effect or a geometrical effect on the living cell. Actually, it was already known from the work of T. H. Morgan (1895) that if one of the first two cells is alive and the other dead, all one has to do to undo the half pattern and institute a whole pattern is to invert the preparation.

More is now known about the interaction of yolk and non-yolk in the establishment of pattern, but before considering it, the role the sperm plays in pattern formation should be considered. A number of investigators (e.g. W. Roux, 1887; J. W. Jenkinson, 1909; M. Herlant, 1911) demonstrated that a sperm can enter an amphibian egg at any point close to the equator and that the mid-dorsal meridian, which we now know as the meridian from which organization spreads, will be 180° away. One way this was learned was by applying a sperm suspension locally to a "dry" egg and then observing the subsequent development.

Many very able people repeated this experiment, could not confirm the results, and came to believe that there is no relationship between the entry point of the sperm and the dorso-ventral axis. Now we know that the sperm can determine the dorso-ventral axis, but its effect can be undone very easily by subsequent treatment. Just as the pattern of a half egg can be changed to that of a whole egg by changing its position, so too can the pattern determined by the entrance point of a sperm be undone by a subsequent change in position of the egg.

An important step toward understanding was taken by P. Ancel and P. Vintemberger (1947). They demonstrated that the plane in which an egg is tipped or turned over becomes the dorso-ventral plane.

How the entry of the sperm and tipping work through a common pathway was explained by Jean Pasteels (1946, 1948, 1964). First consider, as he did, what happens when the sperm enters. He had placed spots of nontoxic dye at a number of positions on the surface of the egg. Shortly after sperm entry, the colored cortical marks on the far side of the egg moved closer to the north pole. Let us think of the egg as the earth and have the sperm enter near New Zealand. If the earth were like an egg, its crust would shift as a result. The greatest shifting would be in the meridian 180° away. The center of South Africa might move up to where the Congo had been. Just lateral to the meridian of greatest northward movement there would be considerable, but less, northward shifting. There would be less and less northward shifting with westward and eastward progression from the Congo–South Africa meridian. Finally points would be reached in the Atlantic and Indian Oceans where no shifting would occur. To carry the analogy further: as the crust shifted, it would carry along with it some of the adhering molten material, causing a mixing of southern and more northern materials. This locus of mixing would be a crescent with its center in the Congo and its two ends 90° away to the east and west.

We return now to the egg. After the sperm has entered, a crescent does form with its center below the equator and on the meridian 180° away from the point of entry of the sperm. This is known as the gray crescent, and its center marks the position of the future dorsal center of organization. According to Pasteels, the grayness results from our seeing some of the southern cytoplasm still adherent to the cortex, pulled up into the less yolky region farther north. What makes this credible is that one can make a new crescent by simply tipping the egg in another plane.

Suppose now that you are holding an imaginary large egg sphere in your lap with its north pole uppermost. A sperm has entered on the mid-left near the equator and has caused a crescent to form on the mid-right below the equator. Now you tip the north pole down and away from you about 15°. If this were a clear glass sphere partially filled with a colored sticky material and above it floated a colorless solution, you would now see above the solidly colored southern region a crescent formed by the sticky material still adhering to the glass where it had been tipped up on the side toward you.

Actually one can do this with a fertilized egg and see the sperm-determined crescent partially obliterated and a new one appear where the egg was tipped up. A line extending through the middle of the crescent and through the poles becomes the mid-dorsal line of the embryo.

When amphibian eggs are laid naturally, they fall by chance with their north poles in any position. They are glued to a substrate and/or to each other, and they are held in that position for some minutes. Later fluid collects between egg and vitelline membrane, releasing the egg to float with its north pole uppermost. The position in which it lay determines the plane through which it rotates as the north pole floats up. Along that plane of rotation, on the upper side, will form the first embryonic organs and from there organization will spread.

It is not known how the axes of all organisms are determined, but where best known, in *Fucus* and in *Amphibia*, the first steps in creating organization are very simple. Pasteels has suggested that the significant step common to the establishment of both sperm-induced and tipping-induced axes is the creation of a region of greatest mixing. Possibly something as simple as the greater mixing of mitochondria and substrates 'gives that region a slight rate advantage in the race toward formation of dorsal structures. Things done along all meridians, such as cell division (Lester and Lucena Barth, 1954) and RNA production, proceed slightly more rapidly along this meridian than along others. When gastrulation begins, it begins dorsally. There in the dorsal lip region glycogen is first utilized (Lucena Jaeger, 1945). Up until the beginning of gastrulation, all regional differences appear to be quantitative. Then, as the result of gastrulation, a new relationship between mesoderm and ectoderm is established along the dorsal side. As a consequence, the first embryonic organs begin to differentiate in that region and the process spreads.

Albert Dalcq and Jean Pasteels (1937) went on from the above studies by Pasteels to consider the implication of these with respect to other knowledge and presented a new theory of development. The essence of it is that through interaction of a yolk gradient and cortex a symmetrical field is established in the cortex, and it is at stations in this global field that the various future developmental events take place.

Students of the problem of differentiation in the amphibian egg have realized for some years that the dorsal meridian may be special only because things done along a meridian are done sooner there. If this is so, one should be able to make another meridian the dorsal meridian by increasing its temperature slightly relative to other regions. This was attempted in the past without success. It has now been accomplished by Richard Glade, Evelyn Burrill, and Richard Falk (1967). The reason for the early failures was that at the time it was not realized that the pattern begins to set before first cleavage. This was also demonstrated by Pasteels (1946). The pattern induced by a tipping of the egg can be undone by a subsequent tipping up until some time before first cleavage. Then subsequent tippings begin to yield dorsal regions between the position of the first and second crescents. Finally, shortly before the cleavage, second tippings have no effect. As noted above, a more massive redistribution as when an egg is inverted may be effective in changing pattern during first cleavage. Glade and his coworkers used the knowledge that the pattern begins to set before first cleavage. They started their eggs in temperature gradients of approximately 2°C across the egg shortly after fertilization. The plane of organization was shifted to the warmest quadrant in a significant number of cases.

Here in the amphibian egg as in a stem of *Tubularia* (Chapter Seven) an axis along which differentiation will occur can form with respect to a difference in temperature. In both cases, a temperature gradient is not the natural determining factor, but it has been used to override the natural factors determining metabolic differences. A consequence of the initial differences seems to be the setting of a physical pattern in the cortex of the egg.

A. S. G. Curtis (1960, 1962a) provided a demonstration of an axis-determining property of the cortex by transplanting pieces of cortex before and after the time Pasteels had reasoned the "setting" must occur. Curtis used the egg of *Xenopus*, the South African clawed toad, and moved small pieces of cortex from one egg to another. In one experiment, bits of dorsal cortex from fertilized but still uncleaved eggs were transplanted to the ventral region of other eggs during first cleavage. The transplanted dorsal cortex from the gray crescent region and the dorsal part of the host both served as centers of organization, and two embryos developed from the same egg.

If such a graft had not been made, the ventral part of the host egg would have formed, after many cell divisions, ventral parts properly aligned with the dorsal structures. A tiny graft, approximately 5 percent of just the surface of the egg, controlled the polarity of development of a large area of the egg. What was transplanted was a bit of cortex and some adhering cytoplasm without the egg nucleus. It had, as far as is known, no special genetic information. When specialized cells later differentiate, they differentiate in a field whose center is established before first cleavage. This field can determine the directions along which control of differentiation will operate.

Curtis also grafted pieces of ventral and north polar cortex between eggs about to cleave for the first time. If either a ventral or north polar piece was placed in the gray crescent region, the integrated system leading to a single embryo was disrupted. Instead, the dorsal regions on both sides of the foreign implant became centers of organization, and two parallel axes close together resulted. This reminds one of the foreign transplants which cause doubling in the eye (Chapter Two). Also in the eye and here in the cortex of the egg, simple cutting does not disrupt the pattern. The cuts heal and the single pattern remains.

Cortical transplants from unfertilized eggs to the dorsal region of eggs in first cleavage blocked development. It is quite clear now that part of the cortex does change at about the time of first cleavage in such a way that it can serve as the center of organization.

There is also a global change which occurs, between first cleavage and third cleavage. The critical time when grafts could change the pattern of organization was close to the time of first cleavage. When Curtis moved bits of cortex around in the 8-cell stage, there was no significant effect. It appears that once a cortical field is established, the small cortical transplants cannot change it. A dorsal graft in the ventral part of an 8-cell stage does not change polarity or regionality. There has been a change in the ventral region so that it no longer responds. The dorsal region of an 8-cell stage still has the power to change the direction of future differentiation in a ventral region of an uncleaved egg.

Now we know that a pattern in the cortex of various invertebrate eggs, in the amphibian egg and in protozoa, is the pattern on which various organs and organelles will appear. In the case of the eggs, the axes of control of differentiation have been established before first cleavage. They are only axes and the various structures which are to appear exist only in coded form at egg time. We now turn to the study of regional interaction along these axes leading to differentiation.

The most complete studies of interaction along axes were performed by Sven Hörstadius on eggs of sea urchins. These and the works of others were summarized by Hörstadius (1939) in one of the great classics of embryology. It is still a must for anyone trying to understand cellular differentiation.

Although the eggs of *Paracentrotus*, the genus on which most of the experiments were made, are tiny, about 1/10 of a millimeter in diameter, their cells can be separated with glass needles under a dissecting microscope. The great advantage is that the cells recombine readily, and regional interactions can be studied.

It was already known from the work of Driesch that the first four cells are totipotent. The first two cleavage planes are meridional and each of the first four cells contains a complete N–S axis. The third cleavage is latitudinal and passes close to the equator, dividing the egg into four northern and four southern cells. Each group has approximately half of a N–S axis. These groups are not totipotent and develop into very different structures, but before going into that we must learn what the various egg regions become as development proceeds. This can be seen in Fig. 8-2. The ring of cells which we call A, around the north pole, spreads out to form about half of the surface of the embryo. Layer B forms most of the rest of the surface. As this is happening, the south polar cells move to the inner part of the sphere. E produces skeleton and some other internal structures. D becomes most of the gut and C forms posterior part of the gut and some of the external region surrounding it.

When Hörstadius separated the four northern from the four southern cells in the 8-cell stage, he was separating very close to a line which would

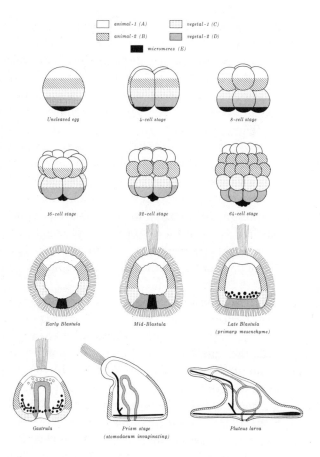

Fig. 8-2. Potency map of regions of the egg of the sea urchin [*adapted from Sven Hörstadius (1939)*].

By applying spots of dye to various regions of the egg, Hörstadius could follow them and learn what the various egg regions become.

come to lie between external and internal cells at the anus in the pluteus larva. Let us consider the four northern cells first. They were presumptive external surface tissue including apical tuft and the early mouth or stomadeum. A peculiar thing was observed. Instead of just a few of the descendants of these northern isolates forming apical cilia, most of them did. Many more cells were given the opportunity to form the most anterior structure than ever do when they are part of the whole. None formed stomadeum.

The four southern cells of the 8-cell stage ordinarily end up as internal cells, mostly gut; however, when isolated, they have nothing to invaginate into and develop rather poorly, but they do form a structure which appears to be gut.

A new relationship was discovered when Hörstadius studied the development of isolated layers and made various combinations of northern and southern cells. The A layer alone made nothing but apical tuft. All—or practically all—of the descendant cells produced the apical cilia. Each was allowed to be the top figure on the totem pole. When layer A was combined with layer E, a very small layer, a tiny but complete pluteus could develop. In this combination of presumptive skin and skeleton, all structures including an intestine developed. No presumptive intestine had been included. The remaining part of the egg, i.e. B C D could also become a perfect pluteus. Here then were two embryos from one egg. The one arose from a combination of polar areas and the other from temperate and tropical regions.

There is a quantitative relationship here. Four E cells in combination with 8 A cells make a perfect balance. Only 2 E cells are required to balance the 8 B cells. If more E cells are added, the larva will have an intestine too large to fit up inside the ectoderm. Recall that this combination of B and E contains no presumptive intestinal material, yet with too much E it can produce more than the normal amount of intestine. Hörstadius produced an overwhelming amount of evidence for this balance concept.

John Runnström (1933) who directed this doctoral work of Hörstadius suggested a way in which the system might work. It was clear that a balance of northern and southern regions was required. Either when alone or when there was too much of one, the system was unbalanced. No region was necessary. All were expendable. Runnström's idea was that differentiation proceeded along a double gradient system. A decreasing concentration of something extended down from the north pole, and another gradient of something else had its high point at the south pole. The high region along one gradient could balance the high of the other. Thus north and south polar regions were in balance and could produce an embryo. Also, the weaker equatorial portions of the two gradients could balance each other and cooperate to produce an embryo. Professor Runnström (1967) now conceives of two gradients of substances which interact and according to their proportions call into action different portions of the DNA. As stated, the substance gradients would be acting directly on the differentiation process. Just about everyone agrees that the idea of a N–S gradient, with bipolar substance differences, is required by the observations. It also seems quite clear that this *leads* to differential DNA activity. To a student of regeneration there appears to be a need for another step.

We know from studies of regeneration that whenever a polarized axis is established, the agent setting up the axis, for example, an open perisarc (Chapter Seven) or a deviated nerve (Chapters Three and Six), does not give information controlling differentiation. Instead it establishes an anterior or distal end and thereby routes along which the cells can interact. The cells lying in this field can then become the most anterior or distal part not lying ahead of them. The theory derived from studies of regeneration is that the initial quantatitive differences measured as rates of oxygen consumption or differences in electrical potential are indications that a polarized communication system has been established. If this theory is extended to embryonic development, the substances associated with a gradient system in eggs would not *per se* turn on or turn off particular genes. What the system appears to do is act as a communication system. As a result of this system in operation, information could flow by which genes in different cells would be controlled. The result in eggs would be, as during regeneration in adults, that each cell would use its most "distal genes" not being repressed.

Many agents are known which can convert a part of an egg into a whole. Many substances have been found which can shift sea urchin development more toward northern development or more toward the production of southern structures (beginning with C. Herbst, 1892 and summarized by Silvio Ranzi and Paolo Citterio, 1954, and by John Runnström, 1967). If, for example, the northern four cells forming the northern hemisphere of an 8-cell stage are simply isolated, many more of the descendant cells produce apical tuft. Even the stomodeum, a structure which normally forms in the north temperate zone, fails to form. All southern structures such as intestine and skeleton do not appear. In fact, these northern hemispheres never gastrulate and live on their stored substrates as permanent blastulae until they die. A number of treatments of these northern halves causes them to develop all structures, sometimes perfectly arranged and proportioned. This can occur for example when the halves are cultured with lithium ion added to the sea water or when tyrosine is added (Mastrangelo, 1966).

Southern halves can be made to produce more northern structures by treatments with thiocyanate ion and other substances, as shown in Table 8-1.

It is difficult to conceive of so diverse and in most cases such simple substances as having the ability to cover or uncover different parts of the genetic code. This is what they would have to do if they controlled genes

Vegetalizing	Animalizing
LiCl	NaSCN
Na$_2$SO$_4$−0.25M	NaI
MgCl$_2$	NaSo$_4$ 0.03M
valine	methylene blue
leucine	lysine
NaN$_3$	lactate
thiourea	maleinate
pyruvate−0.1M	iodosobenzoate
	glucose 0.03M

Table 8-1 (from Ranzi and Citterio, 1954)

directly. It is much easier to believe that the substances which cause regulation toward more complete development act to produce sufficient N–S differences so that there can be communication between cells. The same information is coded in all cells of an individual and, as shown in the many regeneration studies, they will use different parts of the code if they are communicating.

Tryggve Gustafson and Pierre Lenicque (1952) pointed out that in a normal embryo the few apical tuft cells may be able to prevent their neighbors from becoming apical tuft. The fact that they do not in northern halves may indicate a failure in repressive control of the type found during regeneration. According to the theory derived from regeneration studies the animal hemisphere forms too much apical tuft not because of too much apical tuft gene activation but because the transport system is too weak.

There have been many studies indicating a great many measurable effects of the animalizing and vegetalizing agents (rev. by Börje Markman, 1967 and by John Runnström, 1967). These effects are diverse and not in any known way linked to the activation or repression of specific genes. One study stands out because it shows that in addition to affecting N–S development all of the agents have something else in common. Silvio Ranzi and Paolo Citterio (1954) learned that the various animalizing agents listed above all decrease the viscosity of a protein solution whereas the vegetalizing substances all increase viscosity.

The fact that all of so many substances which animalize change viscosity in the same direction and all of the others which vegetalize change viscosity in the other direction makes it virtually certain that the two sets of phenomena, viscosity-fluidity and animalization-vegetalization, are related. One thing to be expected with a change in viscosity is a change in electrical charge. This may be the common action of the various agents. It is something which should be studied.

Lithium has a similar action on development in other eggs. For example, it intereferes with anterior and dorsal development in amphibian eggs (rev. by Pasteels, 1948). In both the echinoderm and the amphibian eggs, the structures most affected, apical tuft in echinoderms and notochord in the amphibians, are at the high end of gradient systems. The strongest argument against lithium acting directly on the genes which produce anterior structures is that genes coding for anterior structures in organisms as different as sea-urchin and frog should themselves be very different and therefore not directly responsive to the same agent. Instead it would appear that lithium works against one end of a system common to both kinds of eggs.

That a field operates in the sea-urchin egg during development is evident from another experiment by Hörstadius. A meridional half of an early cleavage stage was combined with a northern half. A meridional half has a complete N–S axis, and a northern half has only the northern part of the axis. They were combined with cut face to cut face so that half of the resulting sphere had a complete N–S axis and the other half running at 90° had only the northern half of an axis. The N–S axis of the one became the axis of both, and the development in the purely northern part conformed with that in the meridional half, with the result that a well-proportioned single embryo developed. Here is a situation in which one might expect the system to be overbalanced with northern material. Apparently, the control is not determined simply as an addition problem. The direction of control is involved, and one complete polar axis can overcome a partial one and cause realignment. Again the evidence indicates a field in which the material parts are aligned (Fig. 8-3).

In recent paragraphs we have been dealing with eggs whose alignments can be readily changed and along whose axes interaction of parts is easily demonstrated. Such eggs have been called regulative. There are other eggs which, when studied early in this century, appeared to have non-interacting parts. These are the eggs of annelids and molluscs and a few other invertebrate groups. Because it was believed that the cleavage cells of these eggs each received different organ-forming substances and that each part developed independently, each cell was thought to be a part of a mosaic: a mosaic of noninteracting pieces.

There are similarities and also one major difference between the regulative and the mosaic eggs. Already it has been noted that the axes in these mosaic eggs, as in regulative eggs, are determined by the time of first cleavage.

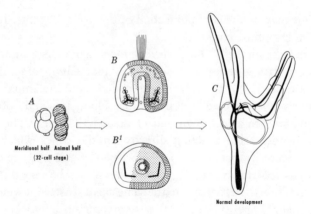

FIG. 8-3. CONTROL OF MERIDIONAL HALF OVER ANIMAL (NORTHERN) HALF IN SEA
URCHIN [*adapted from Sven Hörstadius (1939)*].

a: A meridional half extending from N pole to S pole was fused to a dyed
northern half.

b: The N–S alignment had been determined almost entirely by the meridional
half with a complete N–S axis. Here in the gastrula the apical tuft and the anus,
the most northern and southern structures, are fairly well aligned with the line
between dyed and not dyed tissues. The animal half had shifted the alignment
somewhat, and the entire apical tuft developed from the northern half (cf. Fig.
8-2 for normal development).

b_1: This cross-section through skeletal region of the same gastrula shows that
the dyed northern half had produced more than half of the gastrula. It was
larger to begin with (A).

c: The pluteus which arose from the combination was normal.

It is also quite clear that—as in regulative eggs—there is considerable
redundancy in the sense that more than one embryo can arise from an egg.
Albert Tyler (1930) showed that by subjecting eggs of molluscs and annelids to high or low temperature, anaerobiosis, centrifugation, or pressure,
which realigned the first division spindle, joined twins could be obtained.
When eggs of *Chaetopterus*, a marine annelid, were compressed between
glass slides prior to first cleavage, the cleavage apparatus could be rotated
so that the first two cells, instead of getting quite different cytoplasm and
cortex, each apparently received a whole system and went on to produce a
major part of an embryo. The descendants of the two cells were not completely independent. Some of the structures lying adjacent to each other

made one structure shared by both embryos. This was true for part of the digestive system.

Alex Novikoff (1940) obtained joined twins in the annelid *Sabellaria*. His method was to treat the eggs with KCl added to sea water. A carefully adjusted dosage for a limited time blocked the first cleavage but permitted subsequent development. At the time when second cleavage was expected, some of the treated eggs cleaved into two cells but each had a polar lobe. Each then went on and formed what an isolated polar lobe cell would form, which is essentially a complete trochophore.

This fairly independent development of two embryos side by side and joined after two parallel axes have been established is not peculiar to the mosaic eggs. It is also known for regulative eggs. The case of joined twin formation after frog egg inversion has already been discussed.

Isolates along the N–S axis from mosaic eggs, like those from regulative eggs, also undergo partial development. Some cells, destined to divide only once or twice more when part of the whole, stop dividing at the proper time even when isolated (E. B. Wilson, 1904b). The normal pattern of ciliation may also develop in daughter cells of isolates. However, small N–S fragments develop further when a part of the whole. The notion which has become widespread—that they do alone all that they would have as part of the whole—is wrong. The accounts of all of the many experiments done on mosaic eggs show that N–S isolates produce some of the structures expected. One could never take parts which developed in isolation, put them together, and have the completed larva with all its parts.

Hörstadius (1937), who had earlier done the beautiful work showing the interaction between parts of the regulative echinoderm egg, learned for himself that the parts of the mosaic egg behave very differently when combined. When the northern four cells of the 16-cell stage of the nemertean *Cerebratulus* were combined with the 4 southernmost cells, they did not proceed to form a whole embryo. Instead, the two regions continued to form just what they would have if isolated. In such an experiment the parts appear not to interact. Novikoff (1938) also learned that one sees no effect of grafted cleavage cells of *Sabellaria*. Host and graft went their own ways with no apparent interaction.

Another way it was learned that the two kinds of eggs are different during cleavage stages was by cutting away a few cells. When the most northern cells presumptive for apical tuft were cut away from a sea-urchin egg during cleavage, the most northern cells remaining formed an apical tuft. When the identical operation was performed on the egg of the nemertean, *Cerebratulus*, the analagous structure—an apical spike—did not appear in the young larva (Naohide Yatsu, 1910). However, if one keeps denorthed embryos for 5 to 8 days as did Yatsu and a number of students at the Marine Biological Laboratory, Woods Hole, an apical spike does form. One learns that there is communication between regions such that, when a part is not present, an adjacent region can take over. The difference in the rates of regulation in the two kinds of eggs is remarkable.

Another case of slow regulation was demonstrated by Andreas Penners (1936). He killed a single cell of the 12- or 16-cell stage of the egg of the worm, *Tubifex*. After a number of cell divisions, the descendants of this cell, known as 2d, would normally become nervous system and excretory tubules. When the eggs with the dead cell were kept at room temperature, the young worms developed and lived for a short time without ever having a neural or an excretory system. These structures did appear if the eggs minus their 2d cells were kept at a low temperature. The temperature was low enough to cause an appreciable slowing down of development without completely stopping it. Again we learn that regulation can occur but very slowly. It is interesting to note that the process controlling regulation has a lower temperature coefficient than other developmental processes. This leads to the suggestion that it might be a physical rather than a chemical process.

Axis realignment is also slower in combinations of mosaic eggs than in echinoderm eggs. Recall the experiment by Hörstadius in which he combined a meridional half of a cleaving echinoderm egg with a northern half. The direction of interaction was changed in the northern half and it developed as a meridional half. When Hörstadius made the same combination in *Cerebratulus*, the result was a northern half of an embryo growing out of the side of a half-size whole embryo (Fig. 8-4). The meridional half with a complete N–S axis had regulated to wholeness, but the northern half had developed into what a northern half does as part of a whole, and its axis had not been shifted. The larva in the figure lived for a few days and before its death appeared to be starting to regulate. The long cilia which had arisen from the base of the northern half disappeared. In addition, the two sets of apical spikes appeared to be getting closer to each other.

We learn from this that one part of the regulation process which is slow is the changing of axes. It is as though the parts of a mosaic system were better insulated from each other.

There is evidence that the complete pattern is not stamped out all at once. Anthony Clement (1962), working with the egg of Ilyanassa, a snail, has demonstrated by removing south polar cells at different times during early cleavage that the development of parts such as velum, eyes, foot, and

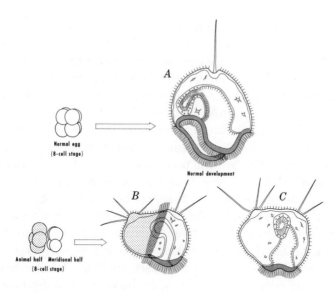

FIG. 8-4.   LESSER CONTROL OF MERIDIONAL HALF OVER ANIMAL HALF IN CERE-
BRATULUS [*adapted from Sven Hörstadius (1937)*].

a: Essentially the same combination of a half N-S axis with a whole N-S axis
at right angles to it was made. In this case there was very little control of one over
the other.

b: Both developed multiple apical spikes and both had the long cilia normally
associated with lappets.

c: A few days later the animal half had lost its lappet cilia and appeared to be
somewhat better aligned with its meridional partner with a complete N-S axis.

shell are better the longer the cells giving rise to them are in contact with
the south polar region and a region north of it which had been part of the
original polar lobe. This indicates an interaction between cells continuing
during early cleavage.

We are left with the realization that interaction does occur during
differentiation in mosaic eggs but that much of it occurs during very early
cleavage and some before the interacting regions are separated by cleavage
planes. Regions of both kinds of eggs, mosaic and regulative, interact before
and after cleavage, but they differ in the relative amount of pre- and post-
cleavage interaction. The greater amount of pre-cleavage interaction in
mosaic eggs has made it possible for them to rely less on interaction between
cells.

Interaction proceeds for a long time in regulative eggs before qualitative differences appear. It is not until the late blastula stage of the amphibian egg that the cells begin to make nucleoli and new ribosomes. Only after that do qualitative regional protein differences begin to appear. Up to the gastrula stage, the differences are quantitative and appear in the form of gradients. During the blastula stage and during much of the gastrula stage, small groups of cells can be transplanted from one region to another without interfering with the pattern of differentiation. Instead of self-differentiating according to their position of origin, they conform to the new neighborhood. For example, a small group of cells taken from a region which would later differentiate as heart can become well-integrated brain tissue if put into the region of the future brain. Reciprocally, presumptive brain can become heart. Everything when moved conforms to the new neighborhood. Only later during the gastrula and neurula stage are the parts determined. Then presumptive heart would not become brain but would continue even in the new neighborhood to self-differentiate partially. It might make a patch of cardiac muscle in the brain. When a part can be removed from its original neighborhood and continues to form something appropriate for that neighborhood, it is said to be determined. Determination is a progressive process and depends upon long continuing interaction. In the course of his exchange transplantations, Hans Spemann (reviewed in 1938) learned that one region is different. It was the same dorsal region, but a day or so older, than when Curtis (1962a,b) transplanted it. In both cases it acted as a center on which a whole embryo formed.

This special region is the one in which visible qualitative differentiation begins. It occurs as gastrulation is being completed. The first cells to take on their characteristic new shape, presumably because of their new special proteins, are cells in the mid-dorsal line. They become the notochord, an elastic rod. Directly above the notochord the mid part of the neural plate develops.

The mid-dorsal mesodermal tissue under normal conditions always becomes notochord. It is distinguished before it becomes notochord by having developed slightly more rapidly than its more ventral mesodermal neighbors. It does not become notochord because it received some special notochord-determining plasm. All of the presumptive notochordal cells can be removed from an amphibian blastula and the most dorsal mesoderm remaining will become notochord. (Klaus Goerttler, 1926; Töndury, 1936; Holtfreter, 1938a). Sometimes as much as the entire dorsal quarter of an amphibian blastula (Curtis, 1962) or an early gastrula of a fish (Luther, 1935) can be removed, and still normal development will ensue.

The dorsal area which later produces notochord is established at the time of first cleavage. The dorsal cortex can be transplanted and become the center of a secondary dorsal field as noted above. It is not specified at the time that the cells developing in that center have to become notochord; what is established is a polarized field. Once the field has been established, the dorsal part of it can be cut away. Then, as in regenerating systems, the most dorsal cells remaining can become notochord.

As late as the early gastrula stage major parts of the amphibian egg can be discarded and the remainder may still form a perfectly proportioned but dwarf embryo. Mabel Paterson (1957) removed large northern and southern areas from young *Rana pipiens* gastrulae. The equatorial third to half egg could regulate to wholeness. The isolating cuts removed the subequatorial center of the dorsal field including the dorsal lip of the blastopore. A new blastopore formed in about 24 hours.

It appears that part of the regulation occurred in the outer surface of the gastrula. If the cut removing the northern third was made slowly so that dark surface cells closed the wound, regulation occurred. When a gap in the original surface was left and lighter internal cells sealed the wound, regulation did not occur.

Curtis (1962) learned that an essential step in regulation occurs a few hours after removal of the dorsal center. If he treated dedorsaled blastulae with the chelating agent, versene, shortly after the operation, they were unable to produce most tissues. If the same treatment was given but not started until three hours after the operation, all tissues could form. Versene binds calcium and can remove it from cell surfaces where it acts in the form of calcium proteinate as a cement. It is suspected that removal of calcium caused the cells to be less firmly bound and therefore to be less able to communicate.

The amphibian egg like the echinoderm egg must have a N–S balance to make or remake a whole system. When Paterson removed just the southern third, the remainder was unable to form a new center. A balanced third could do what an unbalanced two thirds could not.

C. H. Waddington (1938) has shown that at least as much northern tissue as used to unbalance Paterson's equatorial thirds can be added to a whole gastrula without unbalancing it. Waddington grafted a large extra cap of ectoderm to the northern regions of gastrulae. These developed into large but well-proportioned embryos. Apparently in one case where a dorsal area is being established N–S balance is important. After the field has been established, an addition of material just makes each of the organs differentiating in the field larger. The fact that both in the case of equatorial thirds and gastrulae, with extra presumptive ectoderm, the resulting embryos *are* well proportioned tells us that the interactions leading to the differentiation of a part are global. What a particular cell becomes seems to depend upon how many cells there are in the whole egg or how much material there is. For example, a cell lying adjacent to notochord and becoming part of a somite in a small egg would become notochord if sufficient material were added to the system.

As stated above, Spemann learned that one region of a young gastrula does not always conform to its new neighborhood. This is the dorsal area and, as we now know, the dorsal center of a global field. It becomes notochord. If the dorsal area is transplanted to a more northern level close to the north pole, it may conform to its new neighborhood (Töndury, 1936), but if placed roughly at the same latitude on a ventral meridian of another blastula or early gastrula, it may continue to develop as a dorsal center. The graft may invaginate, pass northward under the ectoderm, and differentiate as dorsal mesoderm. At the same time, an almost complete secondary embryo may arise around it.

Spemann suspected that the dorsal graft formed only a small part of the secondary embryo and that it somehow directed the organization of cells around it. To test this, he devised an experiment in which a dorsal graft was taken from a gastrula of one species of salamander and grafted to the ventral subequatorial region of a late blastula of another species. The cells of the two were of different colors. The experiment was performed by Hilde Mangold (Spemann and Mangold, 1924). As one can see in the cross-section (Fig. 8-5), the graft formed notochord and part of a somite. Except for a rather small amount of tissue developing from the graft, all of the rest of the secondary embryo arose from the host. The dorsal mesodermal region came to be known as the organizer.

Before going into the classical organizer concept, let us use the knowledge gained from studies of regeneration and from studies of embryos to construct a new hypothesis. The dorsal region is different as detailed above and, as early as the time of first cleavage, its cortex becomes a transplantable center of global organization. Original polarities can be changed as the result of coming under the influence of this dorsal center. As late as the early gastrula stage, the system is so pliable that a third can become a whole. As late as the gastrula stage, large dorsal parts can be removed, and the remaining most dorsal cells can become notochord.

The new suggestion is that an organizer graft acts first to establish a polarized physical field. Host cells coming under its influence would then receive controlling information coming from a new direction. The graft,

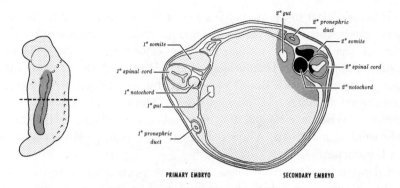

FIG. 8-5. SPEMANN'S ORGANIZER EXPERIMENT [*adapted from Hans Spemann (1938)*].

A second dorsal blastopore region but from an embryo of different pigmentation of another species had been grafted to a young embryo. The graft had invaginated and had extended anteriorly under the darkened area in the drawing on the left.

A cross-section at the level marked by the broken lines is shown on the right. The structures of the host or primary (1°) embryo are at the left. On the right is shown the secondary (2°) embryo. The graft had formed the uniformly dark part: 2° notochord, a small basal part of the 2° spinal cord and part of one of the 2° somites. All of the rest of the 2° embryo had been completed from host tissue. The graft had set in motion processes in the host which led the host to complete the pattern, forming all structures which the graft had not.

being in the high locus of the field, would form mid-dorsal tissues. Each level of the host receiving repressive information from more dorsal levels of the field would form the most dorsal structures not forming in more dorsal regions. Differentiation in an embryo according to this point of view results from a series of repressions. Parts of the hypothesis have been tested in recent years, and we shall return to the experiments generated by the hypothesis.

Now we consider the classic organizer concept. Almost everyone not knowing the results of modern regeneration studies, and presented with the evidence of Fig. 8-5, would naturally think that the cells of the graft were giving specific instructions to the cells of the host. The original idea was that there was a series of qualitatively different inductors. The first ones were thought to originate in the dorsal differentiating tissue. It was assumed that there were different inductors for muscle, cartilage, brain, intestine, and all subdivisions of these.

It became clear in the twenty years after the organizer experiment that during gastrulation, as the mesoderm spreads under the ectoderm, a whole series of changes occur in the overlying ectoderm (rev. by Johannes Holtfreter and Viktor Hamburger, 1955). When ectoderm came into close juxtaposition with dorsal mesoderm, the ectoderm became neural plate and eventually central nervous system. Many such interactions were described. When a differentiating tissue A was combined with undifferentiated X and X→B, A was thought to have induced B-ness by way of a B inductor.

Chemical fractionation of organizer and non-organizer tissue yielded a variety of substances which, when implanted in the ventral part of an amphibian gastrula, would induce a secondary embryo or part of one (rev. by Joseph Needham, 1942). In time, all kinds of substances were found to be inductive. Arthur Cohen (1938) demonstrated that nothing need be added. Electrocautery of a ventral region caused a secondary embryo to form around the damaged spot. After having done much of the work with unnatural inductors, Johannes Holtfreter (1947) came to the conclusion that sublethal injury by a variety of agents could cause ventral presumptive epidermis to become dorsal central nervous tissue. From a vantage point beyond the studies in which morphogenetic polarity was induced by electrical polarity (Chapters Seven and Eight), it seems now that agents which induced secondary embryos may have acted by creating a new center of highest negativity. Injured spots after healing are known to be electronegative to surrounding regions in adult animals. In line with this suggestion were the reports by H. S. Burr (1941) and by H. S. Burr and T. H. Bullock (1941) that the mid-dorsal line of the amphibian blastula is electro-negative to the rest of the surface. It now appears possible that differentiation proceeds in a polarized field and that the great variety of inducing agents all act to create a field. A living graft would in addition add regional information. This is an area where future study should be productive.

Very early in development isolated regions are capable of producing all embryonic tissues. This means that in the early stages many genes are on or that they are easily induced. All parts of the blastula of a fish when cultured under the yolk sac membrane of an older embryo can produce a jumble of embryonic structures such as notochord, striated muscle, neural tube, and digestive tube. As gastrulation proceeds, first ventral, then lateral regions lose the ability to form any specialized tissue (Wolfgang Luther, 1936). There comes a time during gastrulation when only the cells in the relatively small area of the future embryo can form embryonic parts when removed to a yolk sac. This is an important aspect of differentiation. As one set of cells becomes embryo, the parts destined to be extraembryonic become incapable of forming embryonic tissues after simple isolation or transplantation to a yolk sac. They can still be induced to form a part of an embryo by being grafted into a young differentiating embryo. What we are doing when we experimentally induce embryonic tissues is reactivating repressed cells. It is a question whether the cells at the center of organization are ever repressed and whether they ever need to be induced. As in regeneration, so, too, in early embryonic development of regulative eggs, it seems likely that all cells when isolated are capable of forming the dorsal or anterior structures. As part of an organized system, only the most dorsal or most anterior form the structures appropriate to those regions; all others are prevented from doing so.

All eggs are capable of forming several embryos if regions are isolated before repression sets in. This is strikingly apparent when the blastoderm of a duck egg in the blastula stage is cut in the form of a tick-tack-toe field and left in place atop the yolk. As many as 7 complete small embryos can start development (Etienne Wolff and Hubert Lutz, rev. by Wolff, 1948). The point here is that the genes had not yet been repressed and only isolation, not induction was necessary.

During the gastrula stage in the amphibian egg, all cells but those in the center of organization have been repressed. They are reactivated through association with this area. The mid-dorsal ectoderm, when isolated before gastrulation, usually forms a mass of undifferentiated cells. The inductive stimulus is provided by the underlying mesoderm as it glides under the ectoderm toward the north pole. Ectoderm in contact with dorsal mesoderm for a few hours (Holtfreter and Hamburger, 1955) is capable of forming neural tissue.

This reactivation or induction may be a very simple process. Lester Barth (1941) learned that presumptive epidermis from the gastrula of a salamander would sometimes differentiate as neural tissue if simply isolated in culture fluid. The usual thing when this is done with presumptive epidermis of most amphibia is for it to form a ball of cells showing no special characteristics. In these cases something has to happen to the cells before they can begin using their special genes. They can begin to use them after treatment with a number of different ions, for example lithium, sodium, and bicarbonate (Barth and Barth, 1963). Salt treatments cause cultured cells to differentiate into one of at least eight easily recognized cell types. Lithium used for a short time (10 minutes) induces epithelium. The same ion used longer (2 hours) induces nerve tissue and after still longer treatments (5 hours) only pigment cells differentiate. Lester Barth

(1965) thinks that ions may act directly on DNA because there is a close correspondence between the effectiveness of various ions in induction and in their ability to combine with and thus change the electrophoretic mobility of DNA. He has suggested that cations might act by combining with phosphate groups of DNA and thus take sites of repressive histones.

The Barths (1967) are considering the possibility that a single natural inductor acting for different times in different regions might cause the various kinds of cellular differentiation.

It is certainly true that differentiation is controlled by ionic concentration and duration of treatment in these small groups of cells removed from the egg. Here at the forefront of our knowledge, different investigators using different parts of the total knowledge naturally have different hypotheses. My background is such that the first part of the Barth suggestion is welcomed. I believe that the inducible gastrula cells had been repressed as the result of being in a system with a dorsal center and that the treatment with ions derepressed them. Apparently different amounts of treatment activate different stretches of DNA. So far there is no difference in viewpoint.

The difference is that the Barths suggest that differentiation in an embryo might be controlled by ions. I see the possibility that it could be partially controlled by ions. Certainly the first step, when dorsal mesoderm glides up under ectoderm and induces neural plate, could be a process of derepression. The Barth treatments with ions were for a short time. In an embryo this initial action of the mesoderm does not, like the *in vitro* ion treatments, specify the exact direction in which development will occur. Just as in a regenerating system, the activated ectoderm has a number of possibilities open to it. Presumptive spinal cord, presumptive hind brain, and presumptive forebrain, after having been in contact with mesoderm for a short time, all become forebrain when isolated (Eyal-Giladi, 1954). As we shall see later, there are highly specific repressors in embryos. Interactions which follow derepression seem to determine which of the genes will remain active in the specialized cell. I would expect that cells given an ion treatment which would cause them to become epithelial or neural tissue *in vitro* would still develop in accord with their neighborhood if transplanted to an embryo. It would seem that induction is a general derepression and that fine differentiation occurs later. At the moment there are differences of opinion. A clearer view of the process will emerge after more experiments are performed.

One reason why induction studies have been hampered in the past is because investigators have treated several processes as one thing. We speak of the induction of a secondary embryo. This appears to result from the repolarization of a region. Derepression of DNA is also called induction. This apparently opens up large blocks of DNA. Which part of that derepressed DNA will remain active depends on subsequent cellular interaction. That too has been called induction. The acceleration of a developmental process by a steroid hormone is called induction (A. A. Moscona and R. Piddington, 1967). In addition, many people have used induction to mean what the hypothetical inductors do. They are supposed to act by telling adjacent cells what to become. There has been considerable confusion because what appear to be four real and one unreal process have all been treated as one.

Most of the considerable amount of information concerning differentiation was obtained by use of some form of specific induction hypothesis. According to classical induction theory, notochord tells the adjacent cells what to become and this is done by a number of inductors bearing specific turn-on information. It was assumed that after the first tissues had been induced, they in their turn produced new inductors. Eventually, all parts differentiated as they received inductors from their neighbors.

In the head region, for example, there would be inductors for mouth and parts of mouth. Oscar Schotté, working in Hans Spemann's laboratory, attempted to learn something about induction by grafting presumptive belly skin from frog gastrulae to the future mouth region of salamander neurulae. Would the graft form anything at all, and if it did, would the structures be frog or salamander in type? Perfect frog mouths with attendant structures formed in the proper place on faces of salamanders (Fig. 8-6). Spemann (1938) was surprised that anything at all was induced. The great difficulty in understanding this result was that, if one used the inductor hypothesis, the salamander was telling the frog tissue how to make something which the salamander does not have. A frog tadpole's mouth has purely epidermal horny teeth arranged in rows. Near the mouth are suckers which secrete an adhesive. Salamanders do not have this kind of teeth or suckers. If the inductive information were turning on highly specific stretches of DNA, this would require salamanders to have specific derepressors which they would never use. Spemann thought that the salamander was giving relatively unspecific cues, like "form mouth."

From our vantage point today we can see that the salamander need give no specific information to the frog tissue. Isolated frog presumptive belly skin can without any cues produce sucker (Holtfreter, 1938b). This is a case of a tissue in isolation forming a far anterior tissue, something familiar to us in regenerating systems. No isolated piece of undetermined ectoderm

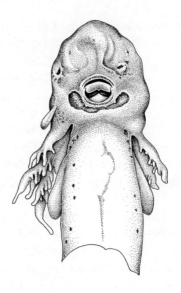

Fig. 8-6.   Frog's mouth on a salamander [*adapted from Hans Spemann (1938)*].

Early in development, before it was determined to form any particular structure, a piece of presumptive abdominal skin from a frog was grafted into the future mouth region of a salamander embryo. The graft transformed to a frog's mouth with horny jaws and rows of closely spaced horny teeth. Just below the mouth is a flattened U of adhesive tissue known as a sucker. All of these structures, horny jaws and teeth and suckers, are peculiar to frog tadpoles. A salamander's mouth is very different (see Fig. 8-7).

would ever form anything as complex as a well-patterned mouth. As noted above, induction involves the production of a field. The grafted presumptive belly skin, when transplanted, was put into an active field. Then not only the sucker but other structures could differentiate because the tissue had been placed in a field where controlling information—its own controlling information—could be transported.

The most striking thing in Schotté's work was that if the graft covered just part of the mouth field, a fairly well integrated mouth, part salamander and part frog, could arise (Fig. 8-7).

At the time there was no escaping the conclusion that the region around the mouth was telling the cells there just where mouth should be. Later Holtfreter (1935) performed a number of these xenoplastic transplantations and learned that such a structure as an ear vesicle could form over much of the head and trunk. This made it difficult to believe that a structure developed in a particular spot because it was told to by a foreign host.

The belief in a general mouth inductor arose only because the graft had been put in a spot at the far anterior end of a system. The same phenomenon is observed in *Tubularia* (Chapter Seven). Any small part of the stem when isolated, if it forms anything, will form mouth and its environs. The comparable situation would arise if we grafted a piece of stem of *Tubularia* to the anterior end of another hydroid and it formed mouth. We know too much about the situation in hydroids to suggest that the host surroundings had induced a mouth. Instead, it is a case of automatically forming the most anterior structure. In both cases, the isolates can become anterior structures. The difference seems to be that the small bit of frog tissue when isolated was incapable of producing the complexity without a field. A sufficient field was supplied by the host.

Most students of differentiation agreed that there had to be some degree of specificity of induction. The experiment which did the most to make me a non-believer was one by Tuneo Yamada (1940). He made various combinations of dorsal and ventral mesoderm from salamander neurulae in ectodermal jackets. A piece of undetermined ectoderm from a gastrula was placed with its inner side up on the bottom of a culture dish. In some cases a piece of differentiating notochord was placed on the ectoderm alongside

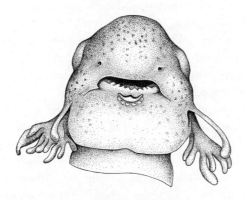

Fig. 8-7. The result of grafting a piece of frog ectoderm over part of the future mouth area [*adapted from Hans Spemann (1938)*].

In this case, most of the tissue forming the mouth is salamander in origin and has made a normal salamander mouth. The goatee-like structure on the center of the lower jaw is a lower jaw of a frog. It arose from presumptive abdominal ectoderm of a frog.

a piece of presumptive kidney tubule tissue. The ectoderm folded up over the mesoderm and formed a ball in which the two pieces of mesoderm are held together. The notochord went on and developed further. The presumptive kidney tissue became striated muscle. This is to be expected if the notochord were telling the adjacent mesoderm what to form. It is proper for somites to arise just lateral to and flanking the notochord. The first specialized tissue to differentiate from them is striated muscle.

The trouble arose when differentiating notochord was combined with presumptive blood in an ectodermal ball. In this case the presumptive blood often became kidney tubules but not muscle. The dorsal-ventral order of the tissues in question is notochord, somitic muscle, kidney, and blood. The fact that the far ventral tissue, presumptive blood, only made it part-way up the ladder under the influence of the notochord indicates that the influence coming from the notochord was something other than a command to make muscle. What do we know for certain that the different-iating notochord does? It certainly can act when transplanted to the ventral part of a blastula to repolarize the environs and become the center of organization. It is the center of a field in which differentiation occurs. Each level, if this is like a regenerating system, would be free to make the most dorsal tissue not forming dorsal to it. Neither presumptive kidney nor presumptive blood would be expected to form more notochord in the small ball where there was already too much notochord for the size of the ball. Both should have been able to produce somitic muscle. Only the more dorsal of the two did. Probably both would have, had there been more time, but the time is limited. Yamada told me that if one starts with already differentiated notochord in the jacket, the adjacent mesoderm does not become more dorsal in its differentiation. This tells us that when the notochord has differentiated to some point, it becomes ineffective. We also know that the lower regions in a field take longer to form the structures proper for the high part of a field. The suggestion here is that with time cut off, the piece can only express a type of differentiation part-way up the ventro-dorsal ladder. This implies that transformation to a more dorsal type of differentiation involves a progression up a linear genetic hierarchy.

This way of interpreting these experiments on embryos is based on information gained from studies of regeneration. It is in agreement with other experiments by Yamada (1950) In these experiments he started with already determined blood. When this far-ventral mesoderm developed in ectodermal jackets, it could become blood cells, and it did not form any of the far-dorsal tissues such as notochord or somitic muscle. The same presumptive, already determined blood, if first bathed in an ammonia solution, could develop into any of the mesodermal tissues of a young embryo. The dorsal tissues, notochord and muscle, were most common and blood formed in only a few of the cases. Yamada wrote of the ammonia treatment as dorsalizing the presumptive blood. Now much later and after the regeneration studies and the studies in molecular genetics, this would appear to be a case of derepressing the entire series of mesodermal genes after all had been repressed except those operating in the far ventral area. This depression could be called induction, but not induction by inductors because the simple ammonia molecule could not carry all of the information for all of the dorsal mesodermal tissues. As in development generally, one expects small isolates capable of forming anything in a series to form the top of the series.

Similar work had been done earlier with isolates of gastrula ectoderm. Lester Barth (1941) had demonstrated that simple isolation of ventral ectoderm from a gastrula of the salamander, *Amblystoma punctatum*, would sometimes enable it to transform to the most dorsal tissue in its series, neural tissue. When ventral ectoderm is simply isolated from gastrulae of most amphibian species, it does not transform to neural tissue. Johannes Holtfreter (1947) demonstrated that a variety of agents enabled ectoderm to transform to neural tissue. What is important to us here is that the tissues formed were not just dorsal but dorsal and anterior. The neural tissue showed resemblances to brain and associated with it were a jumble of olfactory placodes and rudimentary eyes. Some mesenchyme and pigment cells also appeared. Never do well-organized heads arise after these drastic treatments of ventral ectoderm. Dorsal and anterior structures can arise, but the orderly pattern which one sees arising in a normal field is lacking. Instead, there appear to be many rather than one center of organization. For example, as many as 12 olfactory placodes could appear in one explant.

Among the agents which could enable ectoderm to differentiate were solutions of low and high pH, 4.0 and 9.7. R. A. Flickinger (1964) has shown that $CaCl_2$ and guanidine-HCl, in the concentrations which enable ventral regions to transform to more dorsal, were very effective in releasing proteins from yolk platelets. He suggested that the agents acted by breaking hydrogen bonds in the platelets and that the released protein acted to produce dorsal structures. Such agents might be acting as derepressors, as suggested by Barth. The general picture which seems to be emerging is that ventral tissues are already repressed and that a great variety of substances can derepress them. Then as in regenerating organisms a number of developmental pathways are opened up. If a region can become anything, it will become the most anterior and dorsal in its series. It is to be noted that ventral mesoderm when released does not form brain, and ventral ectoderm does not form notochord. The release of these tissues from repression which had relegated them to a ventral role did not make them alike and completely dedifferentiated. Instead, each—ectoderm and mesoderm—had more possibilities in its own series opened up.

The above studies have generated doubt in the existence of information-bearing inductors. Induction is certainly real in the sense that fields of interaction can be induced and one can derepress or induce already re-pressed regions, but the agents such as celloidin beads, NH_3 and Li^+ which do these things do not give specific genetic instructions.

In spite of the fact that no specific inductors are known, it is still difficult to interpret the original induction experiments without them. The experiments were those of Warren Lewis (1904) and Hans Spemann (1905) on lens induction mentioned in Chapter Two. If there are no inductors, is the knowledge of induced fields and repressors sufficient for a reinterpretation of the induction of lens?

We know, as also noted in Chapter Two, that Yo K. Okada and Yoshiki Mikami demonstrated that other tissues than eyecup could be substituted for eyecup and still a lens of sorts would form. This seems to indicate, as we know generally from regeneration studies, that induction consists of two processes. One is the establishment of a field of communication or a repolarization of a former field. Removal of the eyecup presumably stopped the differentiation to lens by breaking lines of communication. Other tissues put into place beneath the presumptive lens position reestablished the contact. The second part of induction, called individuation (Needham, Waddington, Needham, 1934), or simply differentiation, was allowed to continue. Differentiation, as we now know from the work of Jacobson (Chapter Two), proceeds as information arrives from many neighbors, not just the eyecup. The lens was abnormal with a foreign tissue beneath it but was still a lens because its neighbors outside the eyecup region were normal and because the eyecup and the cells from which the lens arose had been in place for a large portion of the pre-visible differentiation period.

When Lewis pushed an eyecup into a more posterior region, it came into contact with ectoderm which would have become part of the skin of the trunk. Before coming into contact with eyecup, the ectoderm had been receiving messages from its neighbors which would have reduced it to the production of a part of the trunk. If this is like the process of differentiation during regeneration, the contact with eyecup would have given it a different stream of repressors. It would be expected that these would prevent it from becoming anything anterior or dorsal to the eye. The former trunk ectoderm would be expected to use its DNA to form any part of the eye not forming upstream. This transformation to a structure of another region can happen after transplantation only when repression has been partial, that is sometime before final determination had occurred. After determination only a treatment such as with NH_3 could have derepressed it so that it could transform.

I favor repressors over inductors as the agents of differentiation because they can be demonstrated. Another reason is that in regenerating systems and in the amphibian egg variously treated, or in some cases untreated, isolates automatically form the most anterior or dorsal structures. More posterior structures arise only when prevented from forming anterior structures. This seems to eliminate a theoretical need for inductors bearing specific turn-on information.

There is another argument favoring repressors. When embryos are differentiating, a particular activity such as lens antigen formation (Chapter Two), or production of cardiac myosin (James Ebert, 1952, 1953), or other heart substances (Howard Holtzer, 1960), or the ability to form a particular tissue (William Muchmore, 1951) is widespread when first observed. As differentiation proceeds, the area of that activity decreases. It is apparent that at least part of differentiation involves turning off parts of the DNA which had been operating in peripheral cells at an earlier stage.

It is still possible that in some cases a product of one cell type acts as an inducer of another cell type. Charles Wilde (1956 and earlier) has presented evidence that phenylalanine produced by dorsal mesoderm may induce melanin production in presumptive pigment cells. This may be a case of induction by a substrate as known in bacteria (Chapter One). Specific induction, as this would be, seems not to be the rule because as noted above, induced cells have more DNA turned on than they will use finally. Differentiation in general is a case of reduction of many potentialities to one actuality. This kind of system seems best controlled by repressors.

The possibility that there might be specific inductors has been studied in a new way by Clifford Grobstein (1962). He starts with organs of mice when there are two components, an epithelial and a non-epithelial part. These can be separated and cultured *in vitro* alone or in combination.

In the case of a salivary gland, one of the components is an inpocketing of the epithelium of the mouth and the other is a condensation of mesenchyme around it. If they remain in contact or if they are recombined after separation, the epithelium becomes a branching tube and the mesenchyme becomes the connective tissue of a respectable salivary gland. Separated, they remain groups of cells which do not form structures.

In the kidney the same holds true for the ureteric bud and the surrounding mesoderm. When together, the bud forms a system of tubules and the mesoderm forms glomeruli. When apart, neither forms structures. From considering this evidence divorced from the general body of knowledge one would suspect that the components were telling each other just what to produce.

The interaction of the components does not require direct cellular contact between cells. The two components can be separated by a millipore filter and still differentiation will occur. There is contact through the pores of the filter by way of cellular products including proteins.

The effect is not specific. In most cases one of the members of a pair can be replaced by mesoderm from another organ and still the epithelium will differentiate typically. The mesodermal components seem to act, at least in part, by producing collagen which precipitates around the epithelium and in some way furnishes a scaffold on which a branching structure can form.

All experiments which have been done to test whether there is specific directive information passed from one tissue to another just before visible differentiation yield the same answer: no. Differentiation as we know begins in vertebrate eggs at the gastrula stage. Potentialities decrease and specification increases. The cells in these mouse rudiments had been differentiating for some days after gastrulation. Obviously they were quite far along the pathway of extreme specification, because they could sometimes take the last steps to their typical specialization in the presence of a foreign partner. The foreign partner does something to enable that last step to be taken, but the effect of their former partners, over the long period since gastrulation, had so specified the cells that they could still complete differentiation typically.

The work with the mouse rudiments by Grobstein and the work of Okada and Mikami in which they substituted other tissues for eyecup appear to be essentially the same. It would appear that the foreign tissue can act with the partner to create a field in which the final steps in visible differentiation can be completed, but that most of the process has occurred before the final step. This belief is strengthened because nerve tissue can be quite effective in trans-filter interaction. It is known from work quoted in other chapters that nerves can set up fields in which differentiation occurs.

One of the best attempts to find a specific inductor has been made by

James Lash (1963). In the vertebrate embryo, cartilage forms around the nerve cord and notochord. Extracts of chick notochord and spinal cord, when added to cultures of chick somites, induce many to produce cartilage. This may be a case of specific induction, but it is impossible to know for certain. Other tissues such as liver also yield extracts which cause cartilage to form, but not as often. The major difficulty in this kind of work is that the isolated somite is very far advanced in differentiation toward cartilage when it is isolated, and it is known from other systems that very unspecific agents such as NH_3 may facilitate the differentiation of dorsal structures. Even the mesonephros farther ventral than the somites, when simply isolated, may form cartilage, especially at its ends. Lash points out that the ends are close to the area where girdle cartilages associated with the limbs develop. This may be a case like that of the lens antigens which seem to be formed over an area broader than the lens prior to the final differentiation. In conformity with general theory, this would appear to be a case where repression of cartilage DNA had not been completed by the time the mesonephros was isolated.

The first test for repressors operating during embryonic differentiation involved culturing eggs from early cleavage till gastrulation along with pieces of adult organs or tissues. Eggs of *Rana pipiens* were cultured in a balanced salt solution either with pieces of adult brain or heart or with blood. The eggs did not react to these treatments in most experiments and developed normally. There were other experiments in which many of the eggs became abnormal during gastrulation and developed with multiple abnormalities. In a few cases, single structures or tissues did not appear on time. In every case when heart was appreciably delayed relative to all of the rest of the embryo, heart had been in the culture medium (Fig. 8-8). Brain could block neural plate formation and adult whole blood blocked blood formation (Fig. 8-9) without affecting the other organs.

The bloodless and heartless embryos had to be kept cool, 9–12°C, because blood flowing through the internal organs is necessary for normal internal development at higher temperatures.

Although the numbers were small, e.g., only 8 purely heartless embryos and 59 with nothing but vascular difficulties, the fact that these appeared only after culture with the homologous tissue made the possibility of this being a chance distribution one in millions.

It was apparent from this early work that although a specifically inhibitory effect was demonstrable, unknown factors were obscuring the results and making further progress difficult. Further work with living tissues and embryos was postponed until the methods for studying the

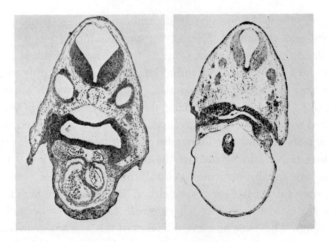

FIG. 8-8.   FROG EMBRYO CULTURED WITH HEART

Both photomicrographs are of cross-sections passing through the hind brain above and the heart below. On the left is a normal animal showing a two-chambered heart with double walls and primitive blood cells in the chambers. On the right is an animal whose presumptive heart tissue has been delayed. It has remained small and the tissue had not become contractile. This and all others like it had been cultured in the same medium with pieces of adult heart.

phenomenon were worked out with worms and hydroids (Chapters Six and Seven).

In the meantime other workers studied the effects of homogenates of embryonic and adult tissues after removal of large cell parts by centrifugation at 3000-5000g. Pierre Lenicque (1959) initiated these studies and learned a number of things from them. He injected his supernatants from blood, brain, eye or heart under the blastoderm of early chick embryos after 18–20 hours of incubation when they were in a primitive streak stage. The embryos were then allowed to develop for 4 more days. As in the work with living frog tissues, many general effects were observed, but against this background of general effects there were some highly specific effects on homologous regions.

Whole blood was quite toxic. After hemoglobin and some of the other proteins were removed by precipitation with ammonium sulfate, most of the general toxicity was removed, allowing the specific effect to show through clearly. Eighty percent of the untreated controls rated 70–80 on the Tallquist scale for hemoglobin, whereas only 20 percent of the blood-treated embryos achieved a rating of 70. Sixty-three percent of the blood treated were at 50 or below on the hemoglobin scale. It is significant that brain-treated embryos and untreated control embryos showed an almost identical range and distribution of hemoglobin scores.

The brain extracts were much less generally toxic than were blood or heart extracts. Brain-treated embryos, however, exhibited easily recognized brain defects in 37 percent of the cases whereas only 4.5 percent of the blood-treated embryos had head defects. The defects caused by materials from brain resulted in the main from a retardation or a cessation of brain development, while other parts of the body continued to develop.

Lenicque's work with eye extracts was particularly instructive. Whereas eye extracts often caused both brain and eye defects, the effect was greater on eyes. Fifty percent of all eye-treated embryos had defective eyes and in some cases the only visibly affected structure was the eye. This eye-limited effect was never observed after treatment with brain. The diameter of a

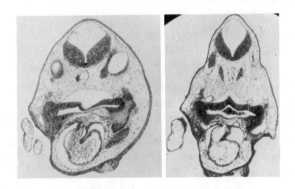

FIG. 8-9. BLOOD-TREATED EMBRYOS

These embryos had been subjected to treatment with blood in the apparatus shown in Fig. 8-11. The one on the left had received negatively charged substances at pH 8.1 from whole blood. Its blood vessels including those in the external gills at the left contain blood cells. The chamber of the heart sectioned at a level which shows a U shape contains many blood cells. There is no apparent effect on blood production of the negatively charged blood substances.

The embryo on the right was treated at the same time in the same apparatus with positively charged blood substances. It had lost cells from the blood-forming area. Its heart and blood vessels contained practically no blood cells. The tissue seen within the chamber is the detached inner lining of the heart.

The method is 100 percent effective with all embryos experiencing disaggregation of cells in the blood-forming areas and a subsequent reduced production of blood cells.

normal eye at an early stage is 1/3 the diameter of the head. Lenicque recognized 3 classes of inhibited eyes. In these the diameters were approximate 1/8, 1/14, and in extreme cases only 1/40 the size of the head. The smallest were really not eyes but just areas where some eye pigment had developed.

It is not surprising that there was overlap in brain and eye effects. These are adjacent regions, and part of the eye forms as an outgrowth from the brain. For a long time during differentiation, adjacent regions seem to be only quantitatively different. There are overlapping patterns of gene-directed activities which later become limited and do not overlap. The case of overlapping lens antigens in embryonic brain was noted in Chapter Two.

At least part of the general effect of blood and heart is not an adjacent region effect. Part of the general effect in these warm-blooded embryos probably results from the fact that moving blood is needed because of its respiratory and excretory function. The results with blood and heart can be more clearly specific in frog embryos because the temperature can be dropped to 10°C. Development can still continue at this low temperature, but the need for blood is minimal.

In Lenicque's work, heart extracts caused the greatest number of general defects. Even precipitation of some of the proteins did not appreciably improve the heart extracts. There were a few heart-treated chick embryos with straight tubular hearts when they should have been chambered and S-shape. The most extreme cases and the most common were mounds of tissues with little resemblance to an embryo. Some of these probably arose as a result of mechanical disturbances arising from the injection beneath the blastoderm because a few similar "embryos" appeared after treatment with extracts of the other tissues. It is significant that although these humps, as Lenicque called them, did not have hearts, only 13 percent contained any pulsating tissue after treatment with heart, whereas 41 percent of the humps arising after treatment with brain pulsated.

Lenicque kept others informed of his progress and helped others get started, with the result that his work was confirmed before his extensive monograph was published. Richard Clarke and David McCallion (1959b) confirmed Lenicque's work with chick embryos. Clarke and McCallion (1959a) also demonstrated that one could specifically inhibit development in the frog embryo by using homogenates. Their method indicated that the effect of the homogenates was during the period after gastrulation during neurulation. This is the time just before and during visible tissue differentiation. In the earlier work with tissues, the tissues had been removed before the neurula stage. It had been noted that material from tissue could remain inside the vitelline membrane. Red color from blood could be observed during neurulation several days after removal of the whole blood. Clarke and McCallion are certainly correct in putting the time of effect after gastrulation.

Maxwell Braverman (1961) and Arthur Katoh (1961) also confirmed Lenicque's work (Fig. 8-10) and went on to discover new relationships. Braverman used different portions of the brain and spinal cord from chicks for his extracts. As Lenicque had shown, the method sometimes yields dead or generally defective embryos, but when the effect was limited to the neural system it could be extremely specific. When Braverman used the far antero-dorsal part of the brain for the extract, that region and only that region of the embryonic brain was blocked. The numbers were: 33 normal, 8 with general body defects and 12 with only the forebrain affected.

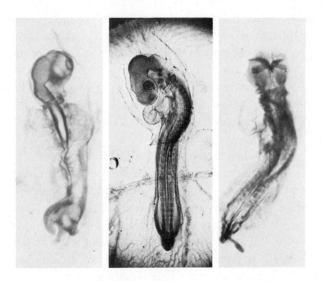

FIG. 8-10.   REGIONALLY SPECIFIC INHIBITION IN THE CHICK [*from Fig. 8 and 9,*
M. H. Braverman, J. Morph. 108: *279*].

The embryo in the center had been incubated for two days and is normal. The embryo of the same age on the right had been treated with a brain extract. The neural plate failed to close, leaving a flat plate of tissue. Eye stalks, optic cups, lenses, heart, spinal cord, and somites appear to be normal, as did all other non-brain structures. The embryo on the left, treated with an extract of blood vessels, had formed a normal brain, but its mesodermal structures are grossly abnormal. The heart developed as two flattened wings of contractile tissue. The somites and the rest of the dorsal mesoderm are amorphous. The spinal cord, whose shape depends upon the form of the dorsal mesoderm, is also abnormal.

There was none with more general neural defects. When more posterior brain regions were used for extracts, there were occasional effects on even more posterior levels, but the rule was that the homologous level and everything anterior to it was affected. Spinal cord extracts quite regularly blocked all spinal-cord development and brain development as well. This is the same relationship which was later observed in *Tubularia*. In addition, when the more ventral brain stem materials were used, they blocked not just ventral development but more dorsal as well.

All regions, according to Braverman, have the genetic information to carry out a series of reactions in which the product of one is the substrate for the next. After differentiation had occurred, the far anterior part would be operating for example with reaction 10. Behind it the next region would be using 9, and in the far posterior region only the first reaction would be in progress. Each region would use the highest-numbered reaction in the series for which it was not receiving repressors. On this basis, the explanation for a hindbrain extract blocking the homologous level and everything anterior to it was that the hindbrain requires reaction 5, for example. The already completed hindbrain had repressors for 5 which would stop the embryo from producing hindbrain. Because there is a sequence with reaction 6 requiring a substrate from 5, and 7 from 6, and so on, all more "anterior" or higher reactions in the sequence would be blocked.

Since the time when Braverman made this suggestion, it has become clear that there are linear series of genes with polarized read-off. (Chapter One). Now such a system should be considered. It seems possible that the series might be a series on a chromosome. If the base sequences coding for posterior structures were at the far right end and those for anterior structures to the left and if transcription ran from right to left, repression at any level might prevent read-off to the left of it, thus preventing the formation of anterior structures. This is for the future.

Braverman used the observed relationship, which he called sequential cumulative inhibition, in a reinterpretation of earlier developmental studies. Marie Tucker (Chapter Six) had learned that substances from more anterior levels of *Lineus* could block head regeneration in more posterior levels. Materials from more posterior levels did not block the transformation of more anterior levels to head. According to Braverman's scheme, if the cut were at a level where, for example, reaction 5 was producing a substrate for 6, it would not be blocked by something preventing reaction $3 \rightarrow 4$, a more posterior reaction, because that had already occurred in the region in question. However, if a more anterior region, such as $7 \rightarrow 8$ were blocked, the region in question could not make it beyond that level of differentiation and would not transform to head. This phenomenon passing through a sequence of ventral to dorsal or posterior to anterior reactions has been referred to as climbing the ladder of differentiation. Finding out what it means in molecular terms is a challenge.

There is a difference in result when one treats a piece of an adult worm and an embryo with materials from a more posterior, already differentiated region. Everything in an embryo anterior to the level from which the material was extracted fails to develop. More posterior materials do not prevent an anterior surface of an adult worm from transforming to head. The difference seems to be that the embryonic region has to pass through a series of steps which the adult region had already passed through. The adult region cannot be stopped by products of more posterior reactions, because it had already achieved a higher, more anterior level of differentiation. Both embryo and adult region can be prevented by anterior materials from becoming more anterior structures.

Braverman also used the theory of sequential cumulative inhibition in an attempt to better understand the works of N. N. Schewtschenko (1936) and Pieter Nieuwkoop (1952). Schewtschenko (Chapter Six) had grafted pieces of either head or tail of planaria at right angles into different levels of whole host planaria. The head pieces grew progressively longer as their sites of implantation were more posterior in the host. The conventional explanation had been that growth is proportional to the differences between the tissues. According to Braverman, each level in the sequence could form all of the distal levels not present at its immediate base. A head in a pre-pharyngeal region could produce everything anterior to that region. A head in a tail region could produce all levels anterior to the tail.

Nieuwkoop had done a similar experiment with amphibian neurulae. Cylinders of ectoderm which would not have differentiated alone were grafted into and at right angles to different levels of the neural plate. There they did differentiate, as we would think now, because they became part of a field in which transport could occur. Each cylinder was able to form all levels of the nervous system down to the level at its base. For example, a graft with its base in the mid-hindbrain region could form fore-, mid- and anterior part of hindbrand. This is the same phenomenon as observed in all regenerating systems. Once differentiation can occur along an axis, all distal parts will form down to the level at the base. It is very unlikely that the base told the ectodermal roll to form forebrain any more than a piece of stem of *Tubularia* tells its distal end to become distal end of a hydranth. It is inherent in living systems that they will form the most anterior structures in their system not already present. They keep doing this by a sequence of inhibitions until all levels are present.

Arthur Katoh (1961) incorporated extracts of chicken tissues in a yolk-agar substrate on which explanted blastoderms could be cultured. As in the works of Lenicque and Braverman, it was learned that brain and mesodermal organs had quite different effects. Brain effects were often limited to the embryonic brain. Materials from mesodermal organs usually left the brain unaffected and had quite different but more general effects. Always, in all of the work on the chick, no matter the source of the mesoderm —dorsal, lateral or ventral—all could interfere with the differentiation of dorsal trunk mesoderm, as in Fig. 8-10. This seems to be a general relationship. More ventral and more posterior materials can block embryonic differentiation not just at the homologous level but also at all higher levels.

Katoh in addition obtained evidence that the control is polarized. Heads of embryos were grafted to anterior ends of whole embryos with the same and with reversed polarity. When they were joined anterior of head to anterior of head, there was no effect. When the posterior end of a head graft was joined to the anterior end of the host, the development of the host's head was inhibited. Graft and host did not remain joined for more than a few hours, and for an effect to be observed the graft had to be appreciably farther along in development—approximately 9 hours more advanced. This is the same relationship—polarized anterior to posterior repression—already noted in all regenerating systems. In such animals as *Stentor* and *Tubularia*, the unions were permanent and the difference in stage between graft and host did not have to be as great.

Norman Dial (1961) studied inhibitory controls in a different way. He combined ectoderm and dorsal mesoderm from frog gastrulae. This combination gave head structures such as brain, eyes, and ears. When the combinations were cultured in extracts of brains or when whole embryonic brains were added to the combinations, the differentiation was shifted toward more posterior brain development and toward more ventral development in the form of pigmented neural crest. This is what one would expect whole brain to do if inhibition is sequential and cumulative. Whole brain consists of all levels, and all of these produce materials which can block their own level plus all more anterior levels. In the event that inhibition was only partial, and it always was in these jumbled combinations, one would expect the anterior regions to be repressed more often because all levels can cause anterior repression but only posterior levels cause posterior repression.

Robert Thompson (1966) has now devised a method for studying lens inhibition which is 100 percent reliable. Heads from chick embryos of 11 to 13 somites, much older than those used in the early studies, were cultured on nutrient media. Extracts of whole eyes added to the media blocked the formation of a vesicular lens in all cases without affecting the formation of the otic vesicle. The neural tissues were only slightly affected, if at all. The success of Thompson's method depends upon the fact, which had already been learned by every student of natural inhibitors, that once a structure has differentiated it is unaffected by the inhibitors. Most of the tissues were already stabilized and widespread effects were not present to mask the specific effect on the lens when high concentrations of inhibitor were used.

It has recently become possible to obtain entirely consistent results with frog embryos. The modern method is based on an earlier observation. Annette Mayfield and S. M. Rose (unpublished), while working with *Rana pipiens* embryos, learned that homogenates of embryonic brain could cause neural plate cells to break and come free from the rest of the embryo. Sometimes the entire neural plate lifted off. Often the effect was limited to the neural plate but sometimes it spread to ventral ectoderm. Underlying mesoderm and endoderm remained intact and alive. The same homogenates of embryonic brain, added to pre-neural plate embryos, greatly delayed the appearance of a neural plate but they did not cause the pre-plate cells to disaggregate. Here then was a substance or substances from embryonic brain which could block the formation of homologous tissue or, if it had already formed, cause its disaggregation and death. This neural disaggregating factor was present from an early plate stage until shortly before differentiation had been completed and the tadpole had begun to feed. Once the neural folds had closed, the embryos were insensitive to the embryonic brain materials.

Charis Shaw (1966) tried to repeat the Mayfield-Rose experiment and was unable to do so at first. It turned out that the vitelline membrane is impermeable to the disaggregating substance or substances throughout most of the year. Only when female frogs had been kept hibernating in the laboratory until a time close to or after their normal breeding period, in March or April, did they give eggs with permeable vitelline membranes. The Mayfield-Rose experiments had been done from March through May. This impermeability probably accounts for the many early failures to demonstrate any effect of tissues on the development of frog embryos (Rose, 1955).

Charis Shaw went on to study the effect of homogenates after vitelline membranes had been removed by trypsin or by watchmakers' forceps. The effect she studied outside the vitelline membrane was an effect not limited to neural tissue but rather to ectoderm generally. Homogenates from embryonic brain and epidermis literally took the skin off embryos. Both neural and non-neural ectoderm disaggregated while mesoderm and endoderm remained intact. This is an effect of embryonic ectoderm on embryonic ectoderm. No other tissue, larval or adult, causes ectodermal disaggregation. Once an embryo has completed most of its differentiation and is close to the feeding stage, homogenates of its skin or brain or any other organ have no disaggregating effect.

The present method of study of organ specific repressors in frog embryos is much the same as those used on *Clymenella* and *Tubularia*. Materials are delivered from organs and tissues to embryos in an electric field. The apparatus is shown in Fig. 8-11. The current passes through a

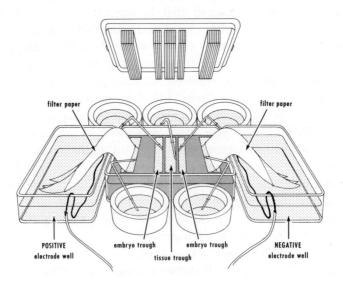

FIG. 8-11. THE APPARATUS USED TO DELIVER CHARGED SUBSTANCES FROM ALREADY DIFFERENTIATED TISSUES TO EMBRYOS

Molten agar is poured into the central chamber, and its cover to which glass plates had been cemented (*above*) is put in place. When the agar is cooled, the cover is withdrawn, leaving 5 troughs in the agar. The two lateral ones contain a balanced salt solution into which the electrode wicks dip. The central trough contains tissue, for example pieces of brain, and the wells adjacent to the tissue well receive embryos. Current passes through the system so that positively charged materials from the tissue reach the embryos in the right trough and negatively charged materials enter the left trough. Embryos shown in Figs, 8-9 and 8-12 were treated in the apparatus.

block of 2 percent agar in a balanced salt solution. A tissue such as pieces of adult frog brain or blood is placed in the central well. Embryos are placed in the nearby lateral wells. The solution used in making up the agar and the solution used in the wells is Holtfreter's solution, a balanced salt solution. The pH of the solution is approximately 8.1. Positively charged substances move from the tissue to the embryos in the right well. Negatively charged substances reach the embryos on the left.

All pre-neural-plate embryos receiving the positively charged substances from adult brain develop to the stage just before plate formation and are held there. Those in the other well receiving negatively charged materials develop normally. If embryos already have a neural plate when subjected to the positively charged materials from adult brain, their neural plates disaggregate while the rest of the ectoderm remains intact (Fig. 8-12). The negatively charged materials at pH 8.1 have no effect.

In the electric field something is removed from adult brain which can pass through vitelline membranes and have its effect. Homogenates of adult brain, which can be quite thick soups, have no effect on embryos in a neural plate stage without the applied field even when the embryos are out of the vitelline. However, homogenates of embryonic brain can stop the formation of a neural plate or cause its disaggregation if already formed. The repressor-disaggregator comes free from embryonic brain but not from adult brain after simple homogenization. It is removed from adult brain by a weak electric field (50 volts over 11 centimeters and a current density of 1.5 milliamperes per square centimeter). It would appear that repressors which are free during embryonic development are loosely bound electrically after differentiation has been completed.

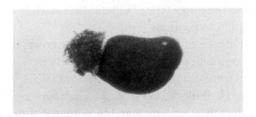

FIG. 8-12.   FROG EMBRYO TREATED WITH POSITIVELY CHARGED MATERIALS FROM ADULT BRAIN

The treatment had caused disaggregation of the brain and its overlying ectoderm. The rest of the embryo remained intact. This loss or more extensive loss of neural tissue is obtained in all cases. The other tissues remain intact and continue to develop for some days.

Materials from blood do not interfere with the formation of neural plate or cause its disaggregation. The positively charged materials from blood have their effect shortly after the neural plate forms. They cause some internal tissue to disaggregate in all embryos, and colorless, non-ectodermal cells can be seen flowing out of the archenteron. They appear to come from the blood-forming area. This has been verified histologically. Approximately 10 percent of the embryos are a day or more late in forming blood. It would appear that other cells can take over after the treatment and produce blood.

Specific repression has been studied in tissue cultures. If embryonic liver cells are grown in a culture medium in which fibroblasts and other kinds of differentiated cells have grown, the embryonic cells may develop either as macrophages or as granulocytes. The effect of the conditioned medium is spoken of as induction (Yasuo Ichikawa, Dov Pluznik and Leo Sachs, 1966, 1967). This is questionable because many substances and generally poor culture conditions had been shown by M. Chèvremont (1948) to cause a number of different kinds of cells to transform to macrophages. This appeared to be very unspecific induction.

Another aspect of this culture work is very clear. If embryonic liver cells are cultured with increasing amounts of differentiated macrophages, the number of macrophages produced by the liver cells decreases to zero and only granulocytes appear. This repression of macrophage differentiation could occur through agar and dialysis membranes. The macrophages were cultured on the bottom of a dish. A layer of agar 1.5 mm thick was poured over them and a dialysis membrane placed on the agar. The embryonic liver cells continued their differentiation above the dialysis membrane. The data clearly indicate that already differentiated macrophages can suppress their own type of differentiation while allowing another type of differentiation to continue.

There is a difference of opinion at present concerning whether there are sufficiently strong bioelectric fields during early development to move charged repressors from cell to cell. Earlier studies (W. A. Dorfman, 1934; H. S. Burr and T. H. Bullock, 1941; R. A. Flickinger and R. W. Blount, 1957) indicated that there are measurable potential differences between regions. Modern measurements with microelectrodes which penetrate cells with a minimum of injury do not indicate potential differences between cells at opposite poles of cleavage stages (Werner Loewenstein, 1966; Shizuo Ito and Nobuaki Hori, 1966). These modern measurements were made with electrodes in the cytoplasm.

Other studies beginning with those of A. L. McAulay, J. M. Ford, and

A. B. Hope (1951) indicate that there are electrical fields surrounding developing systems. This early work demonstrated that fields are present outside plant meristems. Measurements were made between points in the surrounding medium rather than between cytoplasms of different cells. Lionel F. Jaffe (1966) detected extracellular current adjacent to eggs of *Fucus*, an alga. He demonstrated appreciable potential differences with the rhizoid end being electronegative. He believes that the measurement of extracellular currents, since they reveal internal currents, is the superior method. He further believes that there are substantial interpretative difficulties and high noise levels which may detract from the value of studies with intercellular microelectrodes.

One thing which the internal microelectrodes show clearly is that the resistance between cells during early development is several orders of magnitude less than the resistance between the interior of a cell and the external medium. Differentiating cells are electrically coupled (David Potter, E. J. Furshpan, and E. S. Lennox, 1966; Loewenstein, 1966, 1967). Electron micrographs indicate that there is no space between the tight junctions of differentiating cells (Robert Trelstad, Elizabeth Hay and Jean-Paul Revel, 1967). The outer of the two leaflets of the tightly joined cell membranes fuse.

It is often assumed that materials controlling differentiation would have to pass from cytoplasm to cytoplasm. Another possibility is that instead of the communication being across tight junctions from cytoplasm to cytoplasm, it is by way of the fused cell membranes at the surface of the cells. With our present state of knowledge we must keep open the possibility that in tightly joined cells as in the single-celled *Stentor* (Chapter Four), communication during differentiation may be by way of the cortex and that differentiation in the beginning of development is a cortical phenomenon. It is quite possible that cortical differentiation is a fundamental kind of differentiation shared by both unicellular and multicellular organisms. In addition the multicellular forms are able to have differentiation of their nuclei.

We know that the first steps in differentiation in eggs are cortical (earlier part of this chapter). The fluid cytoplasm in either *Stentor* or eggs appears to be irrelevant for differentiation. We should probably be looking at surface potentials as well as at internal potentials. Again we have reached the forefront of our knowledge and there are always differences of opinion in this region.

During early development, the amphibian egg operates as a single field with all cells reading off from the same genetic sequence. Any small group of cells can produce anterior structures if placed in the proper position or if derepressed and isolated. In *Tubularia* it was demonstrated that when anterior parts are combined, some of the cells are forced to transform to more posterior structures. The same kind of experiment was performed with anterior parts of amphibian eggs. G. Lopashov (1935) isolated pieces of presumptive head mesoderm from gastrulae. Single pieces could produce muscle. When more and more pieces of presumptive head mesoderm were combined, more and more tissues arose. His largest masses formed a great variety of tissues, in fact, jumbled embryos. Some formed posterior structures in the form of tail-like outgrowths. In Lopashov's jumbled embryos polarity was not controlled, and one would not expect a unified whole to be formed. It was clear, however, that posterior as well as anterior structures could arise in the larger masses. It is also clear that gastrula cells are alike in that they can all form anything in the whole embryo.

As a consequence of the differentiation which begins during gastrulation, the embryo becomes subdivided into territories such as eye and limb. Each of these territories becomes a sub-field in which different genetic sequences are transcribed. As differentiation proceeds in these sub-fields the same rules apply. For example, all cells of a limb field can use their genes in making any part of the limb. Just which part of the sequence will be used depends upon the position of a cell relative to its neighbors. All levels can transform to more distal levels during regeneration. During early development, as well as later on during regeneration, the control appears to be repressive and to operate from distal to proximal. This appeared in the work of Madeline Kieny (1964). When wing buds and leg buds of chick embryos were combined with the wing distal to the leg bud, the wing differentiated normally but much of the differentiation of the leg was suppressed. When the leg bud was grafted distal to the wing bud, the leg self-differentiated and much of the wing was suppressed.

As the territories become established, the cells of mesoderm and ectoderm are in communication. Edgar Zwilling (1964) has studied this interaction during development of the limb in chick and duck embryos. It is possible by using versene and trypsin to separate the two components, ectoderm and mesoderm, and combine them in various ways. It is quite clear that each is important to the other. Each is becoming specialized in concert with the other. One cannot combine non-limb ectoderm and limb mesoderm just before outgrowth of the limb is expected and still get a limb. The other wrong combination, limb ectoderm and non-limb mesoderm, is also ineffective. By the time differentiation has proceeded to the point of bud formation, only ectoderm and mesoderm from limb buds can cooperate. Wing and leg buds are still enough alike so that the ectoderm from one and mesoderm from the other will cooperate. In general, the type of appendage is determined by the mesoderm.

One might think that wing and limb cells are interchangeable in this very early stage of limb development because each is capable of forming all of the tissues of the other. There is something more than that. When mesoderm of the back is combined with limb ectoderm, a limb bud does not form. The dorsal mesoderm was potentially cartilage and the other connective tissues, muscle, and blood vessels found in a limb. Its failure to produce a limb has been traced to an inability to do something which precedes tissue formation.

A special duty of the distal part of the ectoderm of a limb bud is the formation of an apical ridge. If limb mesoderm is provided with two apical ridges, two poles of limb formation are established and two limbs or two partially fused limbs can arise. The mesoderm is far from being a passive follower. Foreign mesoderm fails because it does not support the formation or maintenance of an ectodermal apical ridge.

As in other developing systems, after an axis of communication has been established, some cells along that axis may be removed without resultant loss of parts of the finished structure. Hansborough (1954) cut away large parts of a limb bud and, provided that he did not remove the tip, all parts of the limb formed. It appears to be generally true that, provided a communication system is in effect, remaining cells can take over the duties of cells removed and complete the whole pattern.

By the time the territories have been established, one does not expect all regions to form anterior parts of the body after something distal to them has been removed. Instead, they tend to produce something more distal in their sub-sequence. This means that once a sequence, such as one operating in the mouth region or the eye, has been turned off, it is not derepressed as readily as is a part of a sequence already being used. Upper arm can readily become finger but we would not expect its mouth or eye-pigment or alpha-crystallin genes to become active during regeneration of the distal part of a limb.

From information such as the above one might conclude that differentiation is a one-way irreversible process. From other experiments it is clear that dedifferentiation can be more complete and that genes for foreign territories can be made operative again after repression. This is clearly the case when cells of a tail or cells of an intestine become well integrated parts of limbs (Chapter Three).

The same thing has been learned about nuclear differentiation by transplanting nuclei. Robert Briggs and Thomas King (1952) removed the nucleus from single-celled eggs of the frog, *Rana pipiens*. A substitute nucleus was injected into the egg. Could a nucleus which had been operating as part of a differentiated cell give rise to descendants so completely derepressed that they could code for a whole embryo? Nuclei from cells of blastulae were totipotent. When they were used as substitutes in an uncleaved egg, they could support total development to the adult stage. This was to be expected because qualitative differentiation does not begin until the very late blastula or early gastrula stage. When nuclei were taken from cells of progressively more advanced gastrulae, neurulae, and tail-bud stages, the stages when differentiation is occurring, fewer and fewer could be found which could support embryonic development (Briggs and King, 1957). Many could support cleavage but development stopped before gastrulation was completed. Finally in older embryos no nuclei which could support embryonic differentiation were found. The nuclei appeared to have become irreversibly specialized and unable to give rise to descendant nuclei which could use a variety of genetic sequences. However, nuclei from much more highly specialized cells of the functional intestine of *Xenopus laevis* tadpoles can support total development from the single-celled egg (J. B. Gurdon, 1962).

In addition, nuclei from a carcinoma of the kidney of adult frogs have been able to carry eggs through to the formation of all embryonic tissues (Thomas King and Robert McKinnell, 1960; Robert McKinnell, 1962; Marie DiBerardino and Thomas King, 1965). A most convincing demonstration of pluripotency of the tumor nuclei has been made by Robert McKinnell, Beverly Deggins, and Deirdre Labat (1969). They induced renal tumors in triploid frogs. Triploids, instead of having the usual two sets of chromosomes, have three sets. The nuclei retain the extra set of chromosomes and their descendants are identifiable. When triploid tumor nuclei were injected into these denucleated eggs and development ensued, all of the tissues were triploid. These embryos, although containing all parts, are abnormal and the best die as young tadpoles. Di Berardino and King have shown that chromosomal abnormalities including breaks appear in tumor nuclei after they have been transplanted to an egg. Nuclei of normal kidney suffer in the same way and are even poorer at supporting embryonic development. J. B. Gurdon (1966) has shown that even though very few brain or red cell-nuclei are making DNA at any one time, 70 percent begin to make it in an hour after the nuclei are placed in a frog's egg. He suggests that the failure of a nucleus to support differentiation may result from slowness in getting started and that chromosomes may be separated before replication is completed. This would cause chromosomal abnormality and subsequent inability to take part in differentiation. The fact that some specialized nuclei can be found which support total development is a clear demonstration that repressed genetic sequences are not irreversibly repressed. The same is true when the progeny of a single specialized cell of a carrot became a whole plant (Chapter Four).

There is no doubt that different parts of the total DNA are used in different cells. Nuclei do differentiate and under the conditions of normal embryonic development they may never be called upon to reuse a genetic sequence after it has been turned off. In this sense differentiation is irreversible. We also know from regeneration studies and from nuclear transplantation that all repressed regions can be made active again. To be able to do this, a nucleus transplanted to an egg must not lag behind in chromosomal replication and its progeny must be ready by the time of gastrulation to begin to use different portions of the DNA which had not been in use in the ancestral transplanted nucleus.

There is another important property of differentiating cells which enables them to form communication systems smaller than the whole egg. This begins to appear very early in development. When an amphibian embryo has become three-layered, ectodermal cells show affinity for ectodermal cells and endodermal for endodermal. When ectoderm and endoderm are mixed, they separate from each other. Mesoderm has affinity for mesoderm and in addition mesodermal cells can make very tight unions with ectoderm or endoderm (Johannes Holtfreter, 1939, and rev. by Johannes Holtfreter and Viktor Hamburger, 1955). Essentially the same is true in the chick embryo, as studied by Elizabeth Hay (1968) with the light and electron microscopes. During the time of mesodermal differentiation after gastrulation, the mesodermal cells lose contact with the other tissues, but the mesodermal cells are tightly joined to each other. One can generalize that as differentiation is occurring and as cells are becoming specialists in a sequence such as the notochord to blood sequence, there are extremely tight contacts all of the way across the system. These close contacts are of the type that allow electrical coupling.

During differentiation, cell surfaces become different and cellular recognition becomes possible. Differentiated or partially differentiated tissues can be dissociated with trypsin and chelating agents until all the cells are separated. The cells of dissociated embryonic kidney and wing buds of chick embryos have been mixed in a random arrangement. The single cells are restless and continue to glide over one another until they have found their own kind. Larger and larger groups are formed as they pick up more and more recruits. Movement continues in the groups of reaggregating cartilage cells from the wing bud until most of the cells are completely surrounded by their own kind. When the system comes to rest, all of the differentiating cartilage cells have come together in groups. They then produce matrix around themselves. The kidney tubule cells also stop when they have met their kind, but they are satisfied and come to rest when all lateral surfaces have made contact and their two ends are free. In this way tubules with walls one cell thick are made (Aron and H. Moscona, 1952; Aron Moscona, 1960; J. P. Trinkaus and M. C. Gross, 1961).

These tissue specific recognition substances at cell surfaces have remained enough alike during vertebrate evolution so that mouse cells and chick cells when mixed, sort not according to animal but according to organ or tissue. Chimaeras of mouse and chick cells of the same tissue are formed as demonstrated by Aron Moscona (1957). It is a rule in evolution that genes for molecules or parts of molecules so important that a change in them would result in disruption of the system are usually eliminated if they mutate. It would seem that these substances are so important to the life of an organism that mutant forms have been selected against.

Another feature of development is that one tissue as it takes shape may control the form of aggregation of another tissue adjacent to it. The shape and thickness of the walls of the neural tube are controlled by the notochord and somites developing adjacent to it (rev. J. Holtfreter and V. Hamburger, 1955). In turn, the spinal cord controls the pattern of aggregation of differentiating cartilage around it. If a spinal cord is transplanted in an upside-down position, the differentiating cartilage attracted to it forms the skeletal casing in conformity with the upside-down spinal cord (Sybil Holtzer, 1956).

These phenomena like the diverse phenomena of derepression and field information have been lumped as embryonic induction. This kind of patterning seems to result from affinities between already differentiating or differentiated like cells and between certain unlike cells. It may also depend upon an attraction between cell products and foreign cells, such as the attraction between collagen from corneal epithelium and nearby mesenchyme (E. D. Hay, 1968; E. D. Hay and J.-P. Revel, 1969).

As a consequence of cells joining hands in particular ways, communicating groups with their shapes determined by their surface affinities are constructed. Such well-assembled groups of communicating cells reach a certain size and stop growing. In the next chapter we shall consider what happens when closely knit integration fails.

References Cited

Ancel, P. and Vintemberger,˙ P. 1947. Recherches sur le déterminisme de la symétrie bilatérale dans l'oeuf des Amphibiens. *Bull. Biol. France Belg. Suppl.* 31: 1–182.

Barth, L. G. 1941. Neural differentiation without organizer. *J. Exptl. Zool.* 87: 371–384.

———. 1965. The nature of action of ions as inductors. *Biol. Bull.* Woods Hole 129: 471–481.

Barth, L. G. and Barth, L. J. 1954. *The Energetics of Development: A Study of Metabolism in the Frog Egg*, 117 pp. New York: Columbia University Press.

———. 1963. The relation between intensity of inductor and type of cellular differentiation of *Rana pipiens* presumptive epidermis. *Biol. Bull.* Woods Hole 124: 125–140.

———. 1967. The uptake of Na-22 during induction in presumptive epidermis cells of the *Rana pipiens* gastrula. *Biol. Bull.* Woods Hole 133: 495–501.

Bonnet, C. 1769. *Palingénésie Philosophique*, Geneva.

Braverman, M. H. 1961. Regional specificity of inhibition within the chick brain. *J. Morphol.* 108: 263–285.

Briggs, R. and King, T. J. 1952. Transplantation of living nuclei from blastula cells into enucleated frogs' eggs. *Proc. Natl. Acad. Sci.* (Wash.) 38: 455–463.

———. 1957. Changes in the nuclei of differentiating endoderm cells as revealed by nuclear transplantation. *J. Morphol.* 100: 269–312.

Burr, H. S. 1941. Field properties of the developing frogs' egg. *Proc. Natl. Acad. Sci.* (Wash.) 27: 276–281.

Burr, H. S. and Bullock, T. H. 1941. Steady state potential differences in the early development of *Amblystoma*. *Yale J. Biol. Med.* 14: 51–57.

Chèvremont, M. 1948. Le systeme histiocytaire ou reticulo-endothelial. *Biol. Rev.* 23: 267–295.

Clarke, R. B. and McCallion, D. J. 1959a. Specific inhibition of neural differentiation in the chick embryo. *Can. J. Zool.* 37: 133–136.

———. 1959b. Specific inhibition of differentiation in the frog embryo by cell-free homogenates of adult tissues. *Can. J. Zool.* 37: 129–131.

Clement, A. C. 1938. The structure and development of centrifuged eggs and egg fragments of *Physa heterostropha*. *J. Exptl. Zool.* 79: 435–460.

———. 1962. Development of *Ilyanassa* following removal of the D macromere at successive cleavage stages. *J. Exptl. Zool.* 149: 193–216.

———. 1968. Development of the vegetal half of the Ilyanassa egg after removal of most of the yolk by centrifugal force, compared with the development of animal halves of similar visible composition. *Dev. Biol.* 17: 165–186.

Cohen, A. 1938. Induction by cauterization in the amphibian egg. *Collecting Net*, Woods Hole 13:87.

Conklin, E. G. 1905a. The organization and cell lineage of the ascidian egg. *J. Acad. Nat. Sci.*, Philadelphia. 13:1–120.

———. 1905b. Mosaic development in ascidian eggs. *J. Exptl. Zool.* 2:145–223.

———. 1931. The development of centrifuged eggs of ascidians. *J. Exptl. Zool.* 60:1–120.

Costello, D. P. 1939. Some effects of centrifuging the eggs of Nudibranchs. *J. Exptl. Zool.* 80:473–499.

Curtis, A. S. G. 1960. Cortical grafting in *Xenopus laevis*. *J. Exptl. Embryol. Morph.* 8:163–173.

———. 1962a. Morphogenetic interactions before gastrulation in the amphibian, *Xenopus laevis*—the cortical field. *J. Exptl. Embryol. Morph.* 10:410–422.

———. 1962b. Morphogenetic interactions before gastrulation in the amphibian, *Xenopus laevis*—regulation in blastulae. *J. Exptl. Embryol. Morph.* 10:451–463.

Dalcq, Albert and Pasteels, J. 1937. Une conception nouvelle des bases physiologiques de la morphogénese. *Arch. Biol.* 48:669–710.

Dial, N. A. 1961. Inhibitory control of neural differentiation in explants of *Rana pipiens* gastrula ectoderm. *J. Morphol.* 108:311–326.

DiBerardino, M. A. and King, T. J. 1965. Transplantation of nuclei from the frog renal adenocarcinoma II. Chromosomal and histologic analysis of tumor nuclear-transplant embryos. *Devel. Biol.* 11:217–242.

Dorfman, W. A. 1934. Electrical polarity of the amphibian egg and its reversal through fertilization. *Protoplasma* 21:245–257.

Driesch, H. 1893. Zur Verlagerung der Blastomeren des Echinideneies. *Anat. Anz.* 8:348–357.

———. 1900. Die isolierten Blastomeren des Echinidenkeimes. Eine Nachprüfung und Erweiterung früherer Untersuchungen. *Arch. Entw-mech. Org.* 10:361–434.

Ebert, J. 1952. Appearance of tissue-specific proteins during development. *Ann. N.Y. Acad. Sci.* 55:67–84.

———. 1953. An analysis of the synthesis and distribution of the contractile protein myosin in the development of the heart. *Proc. Natl. Acad. Sci.* (Wash.) 39:333–344.

Eyal-Giladi, H. 1954. Dynamic aspects of neural induction in amphibia. *Arch. Biol.* 65:180–259.

Flickinger, R. A. 1964. Isotopic evidence for local origin of blastema cells in regenerating planaria. *Exptl. Cell Res.* 34:403–406.

Flickinger, R. A. and Blount, R. W. 1957. The relation of natural and imposed electrical potentials and respiratory gradients to morphogenesis. *Jour. Cell. Comp. Physiol.* 50:403–422.

Glade, R. W., Burrill, E. M., and Falk, R. J. 1967. The influence of a temperature gradient on bilateral symmetry in *Rana pipiens*. *Growth* 31:231–249.

Goerttler, K. 1926. Experimental erzeugte "Spinabifida" und "Ringembryonenbildungen" und ihre Bedeutung für die Entwicklungsphysiologie der Urodelieren. *Zeit. f. Anat. u. Entwick.* 80:283–343.

Grobstein, C. 1962. Interactive processes in cytodifferentiation. *J. Cell. Comp. Physiol.* 60(Suppl. 1): 35–48.

Gurdon, J. B. 1962. Adult frogs derived from the nuclei of single somatic cells. *Devel. Biol.* 4:256–273.

———. 1966. The cytoplasmic control of gene activity. *Endeavour* 25:95–99.

Gustafson, T. and Lenicque, P. 1952. Studies on mitochondria in the developing sea urchin egg. *Exptl. Cell Res.* 3:251–274.

Hansborough, L. A. 1954. Regulation in the wing of the chick embryo. *Anat. Rec.* 120:698–699.

Hay, E. D. 1968. Organization and fine structure of epithelium and mesenchyme in the developing chick embryo. In *Epithelial-mesenchymal Interactions,* ed. R. Fleischmajer and R. Billingham, Williams and Wilkins Co.

Hay, E. D. and Revel, J. P. 1969. Fine structure of the developing avian cornea. In *Monographs in Developmental Biology,* vol. 1, ed. A. Wolsky. pp. 144. New York: Karger Publishing Co.

Herbst, C. 1892. Experimentelle untersuchungen über den Einfluss der veränderfen chemischen Zusammensetzung des umgebenden. Mediums auf die Entwicklung der Tierre. *Z. wiss. Zool.* 55: 446–518.

Herlant, M. 1911. Recherches sur les oeufs di-et trispermiques de grenouilles. *Arch. de Biol.* 26:103–336.

Hertwig, O. 1893. Ueber den Werth der ersten Furchungszellen fur die Organbildung des Embryo. Experimentelle Studien am Frosch und Tritonei. *Arch. mikr. Anat.* 42:662–807.

Holtfreter, J. 1935. Uber das Verhalten von Anurenektoderm in Urodelenkeimen. *Arch. Entw-mech. Org.* 133:427–494.

———. 1938a. Differenzierungspotenzen isolierter Teile der Urodelengastrula. *Arch. Entw-mech. Org.* 138:522–656.

———. 1938b. Differenzierrungspotenzen isolierter Teile der Anurengastrula. *Arch. Entw-mech. Org.* 138:657–738.

———. 1939. Gewebeaffinitat, ein Mittel der embryonalen Formbildung. *Arch. Exptl. Zellf.* 23:169–209.

———. 1943. Properties and functions of the surface coat in amphibian embryos. *J. Exptl. Zool.* 93:251–323.

———. 1947. Neural induction in explants which have passed through a sublethal cytolysis. *J. Exptl. Zool.* 106:197–222.

———. 1949. Phenomena relating to the cell membrane in embryonic processes. *Exptl. Cell. Res.,* Suppl. 1:497–510.

Holtfreter, J. and Hamburger, V. 1955. Embryogenesis: Progressive Differentiation. In *Analysis of Development*, ed. B. H. Willier, P. A. Weiss, V. Hamburger. pp. 230–295. Philadelphia: Saunders.

Holtzer, H. 1960. Aspects of chondrogenesis and myogenesis. In *Synthesis of Molecular and Cellular Structure*, ed. D. Rudnick. pp. 35–87. New York: Ronald.

Holtzer, S. 1956. The inductive activity of the spinal cord in urodele tail regeneration. *J. Morphol.* 99: 1–40.

Hörstadius, S. 1937. Experiments on determination in the early development of *Cerebratulus lacteus. Biol. Bull.* Woods Hole 73: 317–342.

———. 1939. The mechanics of sea urchin development studied by operative methods. *Biol. Revs.* Cambridge 14: 132–179.

Ichikawa, Y., Pluznik, D., and Sachs, L. 1966. *In vitro* control of the development of macrophages and granulocyte colonies. *Proc. Natl. Acad. Sc.* Washington 56: 488–495.

———. 1967. Feedback inhibition of the development of macrophage and granulocyte colonies, I. Inhibition by macrophages. *Proc. Natl. Acad. Sci.* Washington 58: 1480–1486.

Ito, S. and Hori, N. 1966. Electrical characteristics of Triturus egg cells during cleavage. *J. Gen. Physiol.* 49: 1019–1027.

Jacobson, A. G. 1963. The determination and positioning of the nose, lens, and ear. *J. Exptl. Zool.* 154: 273–303.

Jaeger, L. 1945. Glycogen utilization by the amphibian gastrula in relation to invagination and induction. *J. Cell. Comp. Physiol.* 25: 97–120.

Jaffe, L. F. 1966. Electrical currents through the developing *Fucus* egg. *Proc. Natl. Acad. Sci.* Washington 56: 1102–1109.

Jenkinson, J. W. 1909. On the relation between the symmetry of the egg, the symmetry of segmentation, and the symmetry of the embryo in the frog. *Biometrika* 7: 148–209.

Katoh, A. K. 1961. Polarized inhibitory control of differentiation in the early chick embryo, studied *in vitro. J. Morphol.* 108: 355–376.

Kieny, M. 1964. Étude du mécanisme de la régulation dans le développement du bourgeon de member de l'embryon de poulet. I. Régulation des excédents. *Devel. Biol.* 9: 197–229.

King, T. J. and McKinnell, R. G. 1960. An attempt to determine the developmental potentialities of the cancer cell nucleus by means of transplantation. In *Cell Physiology of Neoplasia.* pp. 591–617. Austin, Texas: University of Texas Press.

Lash, J. W. 1963. Tissue interaction and specific metabolic processes: chondrogenic induction and differentiation. In *Cytodifferentiation and Macromolecular Synthesis*, ed. M. Locke. pp. 235–260. New York: Academic Press.

Lenicque, P. 1959. Studies on homologous inhibition in the chick embryo. *Acta Zool.* Stockholm 40: 141–202.

Lewis, W. 1904. Experimental studies on the development of the eye in amphibia. I. On the origin of the lens in Rana palustris. *Am. Jour. Anat.* 3: 505–536.

Lillie, F. R. 1906. Observations and experiments concerning the elementary phenomena of embryonic development in *Chaetopterus*. *J. Exptl. Zool.* 3: 153–268.

Loewenstein, W. R. 1966. Permeability of membrane junctions. Conf. Biol. Membranes: Recent Progress. *Ann. N.Y. Acad. Sci.* 137: 441–472.

———. 1967a. Cell surface membranes in close contact. Role of calcium and magnesium ion. *J. Colloid Interface Sci.* 25: 34–46.

———. 1967b. On the genesis of cellular communications. *Devel. Biol.* 15: 503–520.

Lopashov, G. 1935. Die Entwicklungsleistungen des Gastrula mesoderms in Abhängigkeit von Veränderungen seiner Masse. *Biol. Zentralbl.* 55: 606–615.

Lund, E. J. 1923. Electrical control of organic polarity in the eggs of Fucus. *Bot. Gaz.* 76: 288–301.

Luther, W. 1935. Entwicklungsphysiologische Untersuchungen am Forrellenkeim: Die Rolle des Organisationszentrums bei der Enstehung der Embryonanlage. *Biol. Zentralbl.* 55: 114–137.

———. 1936. Potenzprüfung an isolierten Teilstücken der Forellenkeimscheibe. *Arch. Entw-mech. Org.* 135: 359–383.

Markman, B. 1967. *Studies of nucleocytoplasmic interrelations in early sea urchin development.* The Wenner-Gren Institute for Experimental Biology. Univ. of Stockholm, Sweden. pp. 34.

Mastrangelo, M. F. 1966. Analysis of the vegetalizing action of tyrosine on the sea urchin embryo. *J. Exptl. Zool.* 161: 109–127.

Morgan, T. H. 1895. Half-embryos and whole embryos from one of the first two blastomeres of the frog's egg. *Anat. Anz.* 10: 623–628.

———. 1910. Cytological studies of centrifuged eggs. *J. Exptl. Zool.* 9: 593–656.

Moscona, A. A. 1956. Development of heterotopic combinations of dissociated embryonic chick cells. *Proc. Soc. Exptl. Biol. Med.* 92: 410–416.

———. 1957. The development *in vitro* of chimeric aggregates of dissociated embryonic chick and mouse cells. *Proc. Natl. Acad. Sci.* U.S. 43: 184–194.

———. 1960. Patterns and mechanisms of tissue reconstruction from dissociated cells. In *Developing Cell Systems and their Control.* Chapt. 3, pp. 45–70, ed. D. Rudnick. New York: Ronald.

Moscona, A. A. and Moscona, H. 1952. The dissociation and aggregation of cells from organ rudiments of the early chick embryo. *J. Anat.* 86: 287–301.

Moscona, A. A. and Piddington, R. 1967. Enzyme induction by corticosteroids in embryonic cells: steroid structure and inductive effect. *Science* 158: 496–497.

Muchmore, W. B. 1951. "Differentiation of the trunk mesoderm in *Amblystoma maculatum*". *J. Exptl. Zool.* 118: 137–185.

McAulay, A. L., Ford, J. M., and Hope, A. B. 1951. The distribution of electromotive forces in the neighborhood of apical meristems. *J. Exptl. Biol.* 28:320–331.

McClendon, J. F. 1910. The development of isolated blastomeres of the frog's egg. *Am. Jour. Anat.* 10:425–430.

McKinnell, R. G. 1962. Development of *Rana pipiens* eggs transplanted with Lucké tumor cells. *Am. Zool.* 2:430–431.

McKinnell, R. G., Deggins, B. A., and Labat, D. D. 1969. Transplantation of pluripotential nuclei from triploid frog tumors. *Science* 165: 395–396.

Needham, J. 1934. *A History of Embryology.* Cambridge: Cambridge University Press.

———. 1942. *Biochemistry and Morphogenesis*—reprinted 1950. pp. 785. Cambridge: Cambridge University Press.

Needham, J., Waddington, C. H., and Needham, D. M. 1934. Physico-chemical experiments on the amphibian organizer. *Proc. Royal Soc.* (B) 114:93–122.

Nieuwkoop, P. D. *et al.* 1952. Activation and organization of the central nervous system in Amphibians. Part. I. Induction and activation. Part II. Differentiation. Part III. Synthesis of a new working hypothesis. *J. Exptl. Zool.* 120:1–108.

Novikoff, A. B. 1938. Embryonic determination in the annelid, *Sabellaria vulgaris* I and II. *Biol. Bull.* Woods Hole 74:198–210, 211–234.

———. 1940. Morphogenetic substances or organizers in annelid development. *J. Exptl. Zool.* 85:127–155.

Okada, Y. K. and Mikami, Y. 1937. Inductive effect of tissues other than retina on the presumptive lens epithelium. *Proc. Imp. Acad.* (Tokyo) 13:283–285.

Pasteels, J. 1946. Sur la structure de l'œuf insegmenté d'axolotl et l'origine des prodromes morphogénétique. *Acta Anat.* 2:1–16.

———. 1948. Les bases de la morphogénese chez les vertébrès anamniotes en formation de la structure de l'œuf. *Folia Biotheoretica* 3:83–108.

———. 1964. The morphogenetic role of the cortex of the amphibian egg. *Adv. Morphogenesis.* 3:363–388.

Paterson, M. 1957. Animal-vegetal balance in amphibian development. *J. Exptl. Zool.* 134:183–206.

Peltrera, A. 1940. Le capacité regulative dell'uova di *Aplysia limacina* L. studiate con la centrifugazione e con le reazioni vitali. *Pubbl. Staz. Zool. Napoli* 18:20–49.

Penners, A. 1929. Schultzescher Umdrehungsversuch an ungefurchten Froscheiern. *Arch. Entw-mech. Org.* 116:53–103.

———. 1936. Regulation am Keim von *Tubifex rivulorum* Lam. nach Ausschaltung des ektodermalen Keimstreifs. *Zeit. f. wiss. Zool.* 149:86–130.

Penners, A. and Schleip, W. 1928a. Die Entwicklung der Schultzeschen Doppelbildung aus dem Ei von *Rana fusca.* Teil I–IV. *Zeit. Zool.* 130:305–451.

————. 1928b. Entwicklung der Schultzeschen Doppelbildungen aus dem Ei von *Rana fusca*. Teil V und VI. *Ebenda* 131:1–157.

Potter, D. D., Furshpan, E. J., and Lennox, E. S. 1966. Connections between cells of the developing squid as revealed by electrophysiological methods. *Proc. Natl. Acad. Sci.* Washington 55:328–336.

Ranzi, S. and Citterio, P. 1954. Sul meccanismo di azioe degli ioni che inducono cambiamenti mella precoce determinazione embrionale. *Pubbl. Staz. Zool. Napoli*: 25:201–240.

Raven, Chr. P. 1958. *Morphogenesis: The analysis of molluscan development.* London: Pergamon Press.

Ries, E. and Gersch, M. 1936. Die Zelldifferenzierung und Zellspecialisierung während der Embryonalentwicklung von *Aplysia limacina Pubbl. Staz. Zool. Napoli* 15:223.

Rose, S. M. 1955. Specific inhibition during differentiation. *Ann. N.Y. Acad. Sci.* 60:1136–1153.

Roux, W. 1887. Die Bestimmung der Medianebene des Froschembryo durch die Copulationsrichtung des Eikernes und des Spermakernes. *Arch. Mikr. Anat.* Entwicklungsmech. 29:157–211.

Runnström, J. 1933. Kurze Mitteilung zur physiologie der Determination des Seeigelkeimes. *Arch. Entw-mech. Org.* 129:442–444.

————. 1967. The animalizing action of pretreatment of sea urchin eggs with thiocyanate in calcium-free seawater and its stabilization after fertilization. *Arkiv. für Zool.* 19:251–263.

Schewtschenko, N. N. 1936. Der Einfluss des Korpteiles von hoher physiologischer Aktivität auf den Regenerationsvorgang bei Planarier. *Proc. Inst. Zool. Kiev.* 43–64.

Schultze, O. 1894. Die Künstliche Erzeugung von Doppelbildungen bei Froschlarven mit Hilfe abnormaler Gravitationswirkung *Arch. Entw-mech. Org.* 1:269–305.

Shaw, C. 1966. Ectodermal or presumptive skin disintegration in the frog embryo as a response to specific tissue factor(s). *Devel. Biol.* 13:1–14.

Spek, J. 1930. Zustandsänderungen der Plasmakolloide bei Befrüchtung und Entwicklung des *Nereis-Eies. Protoplasma* 9:370–427.

————. 1934. Über die bipolare Differenzierung der Eizellen von Nereis limbata und *Chaetopterus pergamentaceus. Protoplasma* 21:394–404.

Spemann, H. 1905. Über Linsenbildung nach experimentellen Entfernung der primaren Linsenbildungzellen. *Zool. Anz.* 28:419–432.

————. 1928. Die Entwicklung seitlicher und dorso-ventraler Keimhälften bei verzögerter Kernversorgung. *Zeit. wiss. Zool.* 132:105–134.

————. 1938. *Embryonic Development and Induction.* 401 pp. New Haven, Conn.: Yale University Press.

Spemann, H. and Mangold, H. 1924. Über Induktion von Embryonanlagen

durch Implantation artfremder Organisatoren. *Arch. Entw-mech. Org.*
100:599–638.

Thompson, R. 1966. A study of the inhibitory effects of chick whole eye extract
on the development of the lens of the chick embryo *in vitro*. Can. Jour. of
Zool. 44:661–676.

Töndury, G. 1936. Beiträge zum Problem der Regulation und Induktion. *Arch.
Entw-mech. Org.* 134:1–111.

Trelstad, R. L., Hay, E., and Revel, J.-P. 1967. Cell contact during early morphogenesis in the chick embryo. *Devel. Biol.* 16:78–106.

Trinkaus, J. P. and Gross, M. C. 1961. The use of tritiated thymidine for marking
migrating cells. *Exptl. Res.* 24:52–57.

Tucker, M. 1959. Inhibitory control of regeneration in nemertean worms.
J. Morphol. 105:569–599.

Tyler, A. 1930. Experimental production of double embryos in annelids and
mollusks. *J. Exptl. Zool.* 57:347–407.

Waddington, C. H. 1938. Regulation of amphibian gastrulae with added ectoderm. *J. Exptl. Biol.* 15:377–381.

Whitaker, D. M. 1931. Some observations on the eggs of *Fucus* and upon their
mutual influence in the determination of the developmental axis. *Biol.
Bull.* 61:294–308.

Wilde, C. E. 1956. The urodele neuroepithelium III. The presentation of
phenylalanine to the neutral crest by archenteron roof mesoderm. *J. Exptl.
Zool.* 133:407–440.

Wilson, E. B. 1904a. Experimental studies on germinal localization. I. The
germ regions in the egg of *Dentalium*. *J. Exptl. Zool.* 1:1–72.

———. 1904b. Experimental studies in germinal localization. II. Experiments
on the cleavage-mosaic in *Patella* and *Dentalium*. *J. Exptl. Zool.* 1:197–
268.

Wolff, C. F. 1759. *Theoria Generationis*. Halle, 1759. Zweiter Theil (Entwicklung
der Thiere) pp. 98.

Wolff, E. 1948. La duplication de l'axe embryonnaire et la polyembryonie chez
les vertebres. *Comptes rendus Soc. Biol.* 142:1282–1306.

Yamada, T. 1940. Beeinflussung der Differenzierungsleistung des isolierten
Mesoderms von Molchkeimen durch Zugefugtes Chorda-und Neural
material. *Fol. Anat. Jap.* 19:131–197.

———. 1950. Dorsalization of the ventral marginal zone of the Triturus gastrula. I. Ammonia-treatment of the medio-ventral marginal zone. *Biol.
Bull.* Woods Hole 98:98–121.

Yatsu, N. 1910. Experiments on germinal localization in the egg of *Cerebratulus*.
Jour. Coll. Sci. Imp. Univ. Tokyo. 27:1–37.

Zwilling, E. 1964. Development of fragmented and of dissociated limb bud
mesoderm, *Devel. Biol.* 9:20–37.

CHAPTER NINE

Tumor Formation—A Case of Never-Completed Regeneration

What is the essential difference between normal and neoplastic tissue? Let us suppose a very simple case of normal development. Suppose there were a line of cells and only the one at the head of the line divided and produced daughter cells. After each division, one daughter, the one at the head of the line, would divide and the other would begin to differentiate. This process would yield a normally growing, differentiating organism.

Suppose now that two features were added. If one cell in the line could not receive the growth and differentiation-controlling messages and its descendants inherited that inability, they would constitute a tumor.

The germ of this theory came from Professor Erwin Bünning (1952). He had observed that control was clearly directional in plants. A cell could suppress division of other cells in one line but not in another line at 90°. He suggested that if polarized organization failed, the result would be a tumor. He was not concerned with how cells talk to each other. If divisions occurred in random planes, this indicated a lack of that polarized organization necessary for control.

It is now known after many years of research by many investigators that there are several ways in which a cell can lose its ability to respond to growth-controlling messages. One way is by receiving altered genes transmitted from one generation to another. This was difficult to establish.

It was made possible by inbreeding and selecting for a particular kind of tumor. The pioneers were C. C. Little and Leonell Strong.

One line, the famous C3H line, was started with the offspring from a female mouse with a mammary tumor. Her offspring were crossed brother with sister and the grandchildren with their siblings and so on down through the generations. Whenever possible, only those descendants were used as breeders whose female ancestors had developed mammary tumors. The method is so effective that if a female mouse in the C3H line lives to middle age it is almost certain that she will produce a malignant mammary tumor. More than 90 percent do (Strong, 1935). Selection for other kinds of tumors has also been effective. For example, gastric tumor lines and leukemic lines have been established.

This much indicates that cancer is the result of altered genes. The fact that the tumor arises in a particular organ indicates that the affected genes are involved in cellular differentiation. In a mammary tumor line the affected genes seem to operate only in mature mammary tissue. If full development of the mammary glands is not achieved as in males or in ovariectomized females (Leo Loeb, 1919), the tumors fail to appear. Conversely they can be made to arise in males after estrone injections have caused maturation of the mammary glands (A. G. Lacassagne, 1932).

It took many generations of selective inbreeding to establish the high tumor lines. This means that many genes must cooperate to produce a particular tumor. The complexity is great enough and there is sufficient outbreeding in the human population so that we now understand why cancer is not inherited in the usual sense. Even if one were to receive a complete set of balanced tumor genes from one parent, those from the other parent would almost certainly upset the balance. This too was worked out with inbred lines of mice. If individuals of two tumor lines are crossed, e.g., a mammary line and a gastric line, their offspring have no more tumors than found in an unselected population. The range in types of tumors in descendants of a cross between selected tumor lines shows as great variety as in an unselected population.

The importance of the genetic work is that it has taught us that groups of altered genes may lead to a tumor of a particular character in a single organ. At the same time it has been learned that inherited cancer requires too much precise matching to account for a recognizable fraction of the human cases. Many carefully executed statistical studies also indicate that cancer is not inherited in a practical sense.

There is another way to alter a genome and transform it to a tumor producer. One can sometimes take nucleic acids from a tumor genome and add them to a non-tumor genome and thereby transform it to a tumor-producing genome. Whenever a portion of a genome is transferred, the agent is called a virus.

The work with subcellular tumor agents began with the studies of V. Ellerman and O. Bang (1908). This was followed almost immediately by a search by Peyton Rous and his co-workers for subcellular transmissible tumor agents. By 1914 they had obtained effective agents from three different chicken tumors (Rous and Murphy, 1914). The agents passed through filters fine enough to hold back the smallest bacteria. When the filtrates were injected into recently hatched chicks, they came down with tumors at the injection sites.

Several important things were soon learned about these tumor agents. The first was that they were effective only when injected into chicks of the same flock. Apparently the tumor agents could fit and cooperate only with host cells very much like those from which they had come. Looking back we realize now that whether genetic material is packaged in a sperm or in a virus it must be able to cooperate with the genes of the host cell to produce a tumor. This is true of the pristine agents but they can be changed so that they fit with diverse genomes. This was accomplished by transplanting the tumors from chickens of the original breed to other breeds and to other species of birds. In time the agents lost their narrow specificity and acquired the ability to reside in the cells of many species of birds where they caused tumors (Foulds, 1934).

The first step toward understanding how this change of specificity occurred was made by Francisco Duran-Reynals (1942, 1943). Pieces of Rous Sarcoma I were injected into ducklings. Two kinds of tumors arose. One group arose within a short time of the injection. These were typical Rous Sarcoma I and the cells seemed to be antigenically chicken because they could be transplanted to adult chickens where they would grow. These early tumors would not grow in adult ducks.

The other group of tumors arising in the ducklings after injection of Rous Sarcoma I appeared later, 30 days or longer after injection. These would grow in adult ducks and yielded agents which, although they did not cause typical Rous Sarcomas, did cause tumors in ducks. It appeared that a combination chicken-duck agent had been formed.

The same kind of alteration of tumor agents was observed by Kenyon Tweedell (1955) in frogs. The tumor Tweedell used is a common tumor arising in renal tubules of *Rana pipiens* of the northeast and northcentral United States and adjacent Canada (rev. Rafferty, 1964). Balduin Lucké (1938) after whom this adenocarcinoma was named, had learned that he could not get an agent from tumors of Vermont *Rana pipiens* to cause tumor formation in *Rana pipiens* of New Jersey 450 miles away. Tweedell first extended the range by showing that the Vermont agent did not induce tumors in Wisconsin frogs. However, if a graft of a Vermont tumor regressed in a Wisconsin host, that host might come down with a typical Lucké renal adenocarcinoma some months later.

From these tumors in the Wisconsin frogs' kidneys were obtained agents which caused typical Lucké renal tumors in both Wisconsin and Vermont frogs. It was also possible to obtain Vermont-Wisconsin agents from Wisconsin frogs which as tadpoles had received grafts of Vermont tumor (Tweedell, 1955).

A very close relationship between gene and tumor agent was indicated by further study of the C3H line of mice in which practically all of the mature females developed mammary tumors. There are derived lines of the C3H line in which the frequency of mammary tumor production differs. There are high tumor and low tumor sublines. These are genetically alike in most other respects. Bittner (1936, 1948 and rev. by Moulton, 1945) discovered that milk from mothers of the high line could increase the incidence of mammary tumors in mice of the lower line if they were given the milk as babies. The female mice of the high line clearly developed tumors because of a particular genetic profile. At the same time there were particles in their milk which could cause the same kind of tumor to appear in related mice.

Time after time it has been learned that in most cases only cells of very young animals can be transformed by a virus to tumor cells. This shows up very clearly in the work of Ludwik Gross (1952). He used a strain of mice in which some individuals of some subfamilies develop a characteristic lymphosarcoma. He tried to transfer the tumor by sub-cellular materials from a tumor-bearing individual to individuals of subfamilies which did not produce the tumor. As long as the transfers were made to adult mice no lymphosarcomas arose. Success came only when he injected his tumor materials into foetuses *in utero*. New tumor lines just like the donor line with some individuals developing tumors in most generations were established. It is clear that the new genetic material had been inserted but it had to be done before birth.

Recent work with the Lucké renal adenocarcinoma of frogs again suggests that a virus may be the cause of a particular tumor. The belief that virus may be a transmission agent comes from the fact that the Lucké tumors contain virus in the winter but not in the summer. The virus commonly formed in the winter tumors belongs to the herpes group

(Merle Mizell, 1969). Tumor cells of summer frogs free of the herpes-like virus produce it if kept at a hibernating temperature around 4°C. It is clear that the frogs have the genome to produce the virus at all times of the year, but their genes for the virus production are only on in cold weather. Correlated with the fact that winter tumors contain virus is the observation that only in winter can one find many small young tumors in natural populations (Robert McKinnell, 1969). The growth of new tumors begins in cold weather when virus is being produced. It seems likely to many of the students of the problem that the virus is the causative agent. To others it seems that the virus is a byproduct of the activity of a tumor genome rather than its cause in natural populations.

Eggs and embryos can be infected by the virus and as a result develop the Lucké tumor. Kenyon Tweedell (1969) has shown that eggs can be infected by virus while passing through the body cavity from ovary to oviduct. This would lead one to believe that a tumor-causing viral genome might be passed in nature from one individual to another in the egg stage.

Tweedell (1969) has also demonstrated that if one injects a preparation rich in virus from a renal tumor into the kidney regions of embryos, they may produce the typical renal tumor while still tadpoles. Donald Mulcare (1969) has gone further with the same technique and has demonstrated that cell-free tumor preparations rich in virus can cause parts of kidneys of other species of frogs to transform to typical Lucké-type tumors.

These are clear demonstrations that particular viruses from tumors can cause the same kind of tumor when handled by experts in the laboratory. There is still the question whether or not viruses are involved in the spread of tumors in natural populations. Other work by Tweedell (1969) leads one to suspect that viruses may not be involved. A bit of background is necessary for an understanding of this work by Tweedell.

There are in vertebrate embryos two dorsal ridges which produce kidney tubules. These ridges lie lateral to the somites (Fig. 8-5). The most primitive kidney forms from the anterior part of this ridge and is known as the pronephros. In most vertebrates this is a transitory structure functioning only for a short time. Its place is later taken by the mesonephros which develops from a more posterior level of the ridge. In the highest vertebrates, the mesonephros is replaced by an even more posterior kidney, the metanephros. The frog standing somewhere near the middle of the vertebrate evolutionary series has a pronephros in early stages but uses a mesonephros throughout the rest of its life.

The Lucké renal tumor is a mesonephric tumor. The pronephros has disappeared many months before the tumors arise in adult frogs. When

Tweedell injected the renal virus preparations into embryos, the result differed in two ways. First, tadpoles developed tumors long before metamorphosis and secondly these tumors could arise in the pronephros. Neither renal tumors in tadpoles nor pronephric tumors are known in natural populations.

This leads one to doubt that there is sufficient virus in the environment with the means to infect embryos. If there were, we would expect to find pronephric tumors in tadpoles and Lucké tumors in species other than *Rana pipiens*. The fact that these are not found in nature indicates that the mode of transmission may not be viral.

Instead it appears that given the right genome tumors will appear. Keen Rafferty (rev. 1964) has collected and isolated frogs so that virus would not pass between them. Approximately 40 percent from certain geographic areas produce the Lucké tumor. If Lucké virus is injected into freshly collected *Rana pipiens*, about 40 percent produce renal tumors in a few months. The injected animals produce the tumors sooner on the average than do animals which had not been given the virus. It seems likely that about 40 percent of the frogs have the proper genome for Lucké tumor production. If virus is added before they would have produced tumors spontaneously, the virus can start the tumors prematurely.

Whether the Lucké tumors are transmitted by virus or as part of the inherited genome or in both ways is not yet certain. Whatever the final answer, the case is exceptional. In these *Rana pipiens* there is great genetic uniformity with respect to the ability to accept and grow the Lucké virus. In other kinds of animals such uniformity which makes it possible for tumors to be transmitted is only attained by selective inbreeding. It is difficult to believe this is the case in the frogs because the populations which harbor the tumor are so widespread. Rather it would seem that the genes which make tumors possible late in life may confer fitness in earlier life. This is an area difficult to explore but possibly an important area.

The value of wrong or partially wrong hypotheses is quite apparent in the work on tumors already covered. The two major hypotheses were that cancer is inherited and that it is a viral disease. As noted above, the complexity of a tumor genome is so great that it is not passed intact from generation to generation except in inbred lines or when given to an embryo. Because the human population is so genetically heterogeneous, cancer is not inherited in the usual sense of the word.

It is also established, thanks to the pursuit of the viral hypothesis, that part of a tumor genome—the necessary missing part—may be transmitted as a virus. However, cancer cannot be considered a viral disease in the human population. In mammals the donors and hosts have to be almost identical genetically for part of the genome of one to function in the other. However, as noted in the case of the frogs and chickens, the very narrow specificity requirement can be broadened if tumor tissue is transferred through a series of hosts. This does not happen in the human population. It may have happened in the laboratory with the polyoma virus. This now causes tumors in a variety of organs in mammalian laboratory species (Eddy et al, 1958, 1959). Only in inbred lines could one expect to find cancer operating as a transmissible disease. This has happened in a flock of inbred chickens, where lymphomatosis spread through the flock as an infectious disease (Burmester, 1952).

Although the pursuit of the genetic–viral hypotheses indicated that under normal conditions cancer is not transmissible, it has given us part of the understanding of the nature of cancer. As a byproduct it was established that the change from normal to neoplastic is a genetic change and that it involves genes which operate in one kind of differentiated tissue. We shall return to this subject but first we must ask, "if viruses are not the usual cause in the human population of the transformation of a normal genome to a tumor genome, what is the cause?"

It is now clear that there are many environmental agents which can cause the neoplastic transformation. As early as 1779 Percival Pott suspected that the relatively high incidence of carcinoma of the scrotum in chimney sweeps was caused by trapped soot in their trousers being rubbed into the skin at the crotch. There are now many well-documented correlations between an enviromental factor and a tumor. For example, inveterate pipe smokers sometimes develop a carcinoma where the pipe stem touches the oral epithelium. People who work under the high sun are more likely to develop epidermal tumors than people who do not (Harold Blum, 1948, 1959). An appreciable percentage of workers who painted radium on watch dials and tipped the brushes with their lips came down with tumors of the lower head. The most publicized and massive correlation is between cigarette smoking and lung cancer.

The proof of a causal connection between an environmental agent and cancer was given by Katsusaburo Yamagiwa and Koichi Ichikawa (1918). They painted coal tar on an area of skin of mice, peeled it off and replaced it repeatedly. In about a month the first tumors, some malignant, began to appear at the tarred site. When the tar was applied for several months practically every mouse had a tumor at the site of application.

The tar studies opened up the study of chemical carcinogenesis. Many different kinds of compounds were found to be carcinogenic (rev. by Clayson, 1962).

Another correlation was noted—substances which acted on germ cells to produce heritable mutations could also act on somatic cells to produce tumors. The basic change appears to be in the hereditary materials of the affected cells because their cellular descendants bear the same limited ability to respond to growth-controlling messages. This change would not be expected to pass to a son or daughter. Instead it would be expected to die with the affected individual because its germ cells had not been affected.

There are also physical carcinogens. Various kinds of irradiation are now known to be carcinogenic (Fürth and Lorenz, 1954).

It appears now after many thousand man-years of research that the neoplastic transformations can be effected by a number of different carcinogens. Although these mutagen-carcinogens are always with us, the particular mutations leading to neoplasia are extremely rare. This brings us to the major question, "what is the cellular characteristic determined by these neoplastic mutations?"

We know that it has something to do with cellular differentiation. The knowledge of this goes far back to the early days of the study of chicken tumors. By 1914 the Rous group had three agent-transmitted tumors (Rous and Murphy, 1914). The first tumor—still perpetuated in laboratories today by transplantation and agent transfer—is known as Rous Sarcoma I. It has spindle-shaped cells and soft intercellular material. The agent from it when injected into muscles of young chicks induced the cells which received it to transform to a spindle shape and secrete the same kind of viscous intercellular matrix. Another of the tumors was a disorderly growth of cartilage and bone, a chondro-osteosarcoma. The agent from this transformed cells in muscles to the same kind of abnormal cartilage and bone. The third was a fissured tumor unlike any normal organ. Its agent transformed cells so that they arranged themselves in the same unusual pattern. James B. Murphy (1935), one of the Rous group, made this point:

> An already differentiated cell becoming infected with a growth-stimulating parasite might be expected to retain its essential characters in the malignant state, but with the fowl tumours we have evidence that the complex differentiation of the cells is just as much the result of the agent as is the growth.

This knowledge was not used at the time because some other things were still unknown. These had to be discovered before a synthesis could be made.

Later more information concerning these combination tumor–differentiation agents was obtained. Bits of the Lucké adenocarcinoma of frogs were transplanted to limbs of salamanders. Distal portions of the limbs were then amputated and the regeneration of the stumps with included piece of tumor was studied (Rose and Rose, 1952; Rose, 1952). In one case regeneration proceeded in a left forearm until a blastema had formed. Then there was a pause of several days with failure of the new limb to take shape. The stump back to above the elbow was removed for microscopic study. An unusually large amount of cartilage had already formed. In the meantime, a growth appeared in the elbow region of the other forelimb. Nothing had been done to this limb. The growth turned out to be a tumor of abnormally arranged cartilage.

Because of the possibility that the kidney tumor agent of a frog had been involved in the production of the cartilaginous tumor in a salamander, homogenates of the salamander tumor were injected into frogs. If a high percentage of the hosts developed renal tumors, it would appear that the frog renal agent had caused a cartilage tumor in a salamander. Many of the frogs did develop renal tumors but more interesting is the fact that one of the first frogs to receive the materials from abnormal salamander cartilage developed a malignant skeletal tumor. This furnished the opportunity to run a good test for the presence of the renal tumor agent. Bits of the frog chondro-osteosarcoma were grafted to eyechambers of frogs of a different geographical race too young to develop tumors on their own. The chondro-osteosarcoma took in the eyechambers of only three of these small frogs but all three developed the typical Lucké tumor in their kidneys.

Further tests indicated that particles with the ability to produce either skeletal or renal tumors had become associated. The first time was when a frog renal agent was present during regeneration of cartilage in an adult salamander. Later agents with affinity for two or more tissues were obtained by grafting the Lucké tumor to limbs of young salamanders. It seemed that an abnormal nucleic acid determining a renal tumor might fuse with a nucleic acid determining another kind of tissue (Rose, 1952).

That such unions can be made seemed more likely after the work of James Ebert (1959). Ribosomal extracts were made from cardiac muscle of chickens and from Rous Sarcoma I. The extractions were made from the tissues both separately and when they were together in the same container. The extracted ribosomes and associated materials were then tested for activity on chorioallantoic membranes of chick embryos. When the materials from cardiac muscle and tumor were extracted from a mixture of these tissues, patches of tumor and patches of cardiac-like muscle differentiated from the membrane. When the extractions were made separately and the materials pooled and added to the membranes, no cardiac tissue appeared. It seemed that the normal cardiac-determining agents could not take and direct differentiation in embryonic cells unless they had received something from the tumor cells during their extraction together. It seems possible that the agent may be a modified ribosome or a bit of the cell surface pinching off as an RNA virus.

Returning again to the amphibian tumor work—it was noted a number of times that when a tumor arose as the result of a modified agent it was likely to arise in the same position on both the left and the right sides of the body. For example, if a skeletal tumor arose at the first joint of the third finger of the right hand, the next most likely place for a skeletal tumor to appear was at the same joint on the third finger of the left hand. Tumors appeared at bilaterally identical spots far more often than could have been expected by chance.

The fact that these modified agents acquired a specificity of affinity which can be more refined than the specificity for a tissue or an organ appears significant. It may mean in the case of the joint example that the modified agent had acquired from a cell of one joint a structure which enabled it to fit in the same spot on the other side of the body. The implication is that every small part of the body is using a unique part of the genome transcribed as RNA and that the highly specific tumor agents contain modifications of these highly specific RNAs.

All of the genetic studies and all of the RNA virus studies indicate that the neoplastic change is a change in heritable material used in very limited portions of an organism. We also know that when this change has occurred, there is an abnormal patterning of the cells and a failure to stop growing.

The first clear clue of a step between gene and failure of growth control was that tumor cells do not adhere to each other as tightly as normal cells. This is obvious in the case of the almost formless tumors. These, called anaplastic, are more like groups of loose cells than well-knit tissues. Dale Coman (1944) made the generalization that failure to adhere is a neoplastic property. This he was able to demonstrate by spearing adjacent cells with needles of a micromanipulator and measuring the force required to pull them apart. It took much less force to pull apart tumor cells than cells in their tissue of origin.

It was noted in Chapter Eight that differentiated cells of particular kinds stick together in particular configurations. Even after dissociation they will reaggregate each to its neighbors with some or all of their surfaces adhering. Which surfaces adhere apparently depends upon a molecular configuration at cell surfaces. This in turn determines the shape of the tissue and organ.

Control of the individual cell within the group is dependent upon its close adherence to its neighbors. When we cut away a part of a normal tissue, the remainder may grow but in time growth ceases or an equilibrium of production and destruction is attained. A neoplasm is different as Professor G. W. de P. Nicholson (1950), a very wise student of the problem, recognized. It is as though the neoplasm is constantly beginning to regenerate but it never achieves that normally polarized cohesiveness which is the condition for stopping.

It seems well established that neoplasia is the result of a genetic change expressed as a tissue-specific character which leads to abrogation of normal growth controls. The genetic change can be thought of as a primary cause and the failure to respond to growth control as the final result. There are steps between. How does the genetic change result in non-ending growth?

The genetic change appears to be either at the DNA level or at the RNA level. The highly tissue-specific tumor agents which can direct differentiation are large and appear to control multiple functions. There are also small DNA viruses which can cause neoplasia. These are not so limited either in the tissues in which they can grow or in the species which they can infect.

Leo Sachs (1965), who has worked with the small DNA viruses, reasons that the virus carries the code for a protein which becomes part of the cell surface and that its presence in the cell surface changes the cell-regulatory mechanism. A great wealth of information gained by many investigators lies behind this synthesis by Professor Sachs.

Because these viruses are so small and contain so little DNA, only enough to code for about 1500 amino acids, they must have very few functions. Part of their code must be used to produce their own special coating protein and this leaves enough for very few other functions. One of these functions would be the turning off of the viral protein coat production. Maybe all of the remaining small portion of the code would be required for the cell-surface protein recognized as a transplantation antigen. If this be true, the significant neoplastic mutation, in this case introduced by a virus, would act by producing a single surface antigen. A particular virus always induces the same antigen but all new antigens caused by chemical and physical carcinogens are unique (Prehn, 1965; Klein, 1966; Lappé, 1968).

An understanding of the portion of this theory of carcinogenesis requires some background and information. The first to be dealt with is the turning off of the viral protein-coat production. Viruses have been classified as temperate and intemperate. Both enter host cells because of an affinity between protein coat of the virus and surface of the host cell. Once inside, both intemperate and temperature viruses come apart into a DNA portion and a protein portion. If the DNA replicates and also codes for its own special protein, the viral subunits can be assembled into new virus particles. This may disrupt all normal genetic action and lead to disruption of the host cell with release of the newly made virus particles. This is intemperate viral behavior.

The temperate viruses also enter and separate into DNA and protein. Their DNA then becomes part of a host chromosome and serves as part of the host genome. If it does not produce viral protein, or if it does this rarely, it is known as a temperate virus. It may persist indefinitely as an integral part of a host cell genome without killing the host cell. This is the way of the usual tumor viruses. Once the DNA has become part of the host cell, it may never produce complete virus. The recognition of the dormancy of this part of viral function was explained by André Lwoff (1953). The fact of non-production of virus may require activity of part of the small viral genome and is therefore included in the Sachs theory.

A more detailed explanation of the transplantation-antigen part of the theory is required. If foreign substances, primarily proteins, get into the higher vertebrates, a double immune response is generated. The better-known part of the reaction is the production of antibodies which circulate in the bloodstream. The foreign substances are called antigens. These induce the production of complementary molecules called antibodies. Antibodies, once induced by antigens, can combine with more of the same kind of antigen. If the antigens are parts of cells and vital to them, the antibodies may cause the cells to die.

The other part of the immune response depends upon the recognition by lymphocytes of foreign material. When cells are grafted from one species to another, the lymphocytes of the host infiltrate the graft, associate with, and kill the donor cells. The ability of the lymphocytes to recognize minor genetic differences even within a species is great. This was first recognized by Sewall Wright and Leo Loeb (Loeb, 1945). Guinea pigs inbred for 23 generations were still sufficiently different for grafts from one sibling to another to be invaded by lymphocytes of the host. Only when almost complete or complete genetic identity of donor and host is achieved are grafts not invaded by host lymphocytes.

The substances recognized as foreign by lymphocytes are in cell surfaces. They can also serve as antigens. When antibodies are prepared against them, the antibodies can combine with the surface antigens and interfere with their activity.

Elucidation of a phenomenon known as tumor enhancement has done much to increase our understanding of how these surface proteins function. If one starts with a tumor which is different enough so that it will be killed by the lymphocytes of a host, this tumor may in some cases be made acceptable to the host. This was done by pre-injecting the host-to-be with living or dead material from the donor-to-be. This material to serve as a source of antigens could be the tumor itself or any of a variety of normal tissues from the same donor or from another individual of the same inbred strain. After antibodies had been produced in the host, a piece of the tumor could be grafted and paradoxically it was accepted and could grow (George Snell, 1954; Nathan Kaliss, 1958). Instead of the antibodies stopping the tumor, they had made it acceptable to the host.

The explanation came after Göran Möller (1963) showed that tumor cells treated *in vitro* with antibodies against their own surface antigens and then transplanted had been made acceptable to the host lymphocytes. This seems to mean that lymphocytes cannot recognize surface substances as foreign if they are coated with antibody. This is important because it demonstrates that the surface substances by which cells are recognized are antigenic.

In the work with transplantation antigens, the number of antigens may be large. The more distantly related donor and host are, the more antigenic differences are found. The fact that during enhancement the antibodies against one tissue can coat cells of another tissue indicates that when one deals with transplantation antigens during tumor enhancement, one is dealing with antigens common to many tissues.

When a cell becomes neoplastic, it and its descendants differ from normal cells in their ability to recognize their own kind. There are always new antigens demonstrable in neoplastic cells (R. T. Prehn, 1965; George Klein, 1966). According to the reasoning of Sachs, the neoplastic DNA may code for only one molecular species of surface antigen. The addition of a new cellular surface antigen may be the missing step between the genetic change and the failure of growth regulation.

Could the failure of a cell to unite with its own kind because of the addition of an improper antigen be the critical event? The result of the presence of the new tumor antigen is always the same whatever the cause of the new antigen. This shows up clearly when a few cells in monolayer cultures are transformed to tumor cells. No matter whether the cells are transformed by a virus, or by X-rays or by chemical carcinogens, the affected cells do not remain as part of the tightly-knit sheet of cells. Instead, they come free from the sheet and they and their descendants begin to mound up above the single sheet of normal cells (Carmia Borek and Leo

Sachs, 1966). Associated with the changed architecture in the culture is an increase in cell division.

A changed architecture and an unusual pattern of cell division was observed in the first tumor cultures (Alexis Carrel, 1925). When explants of normal tissues are made on plasma clots they first spread out over the clot. Later mitoses begin to appear near the periphery. Cells lying behind, closer to the center, do not divide. The first tumor cultures differed in that there were gaps caused by death of some of the tumor cells. Correlated with this were mitotic figures scattered throughout the cultures.

A tumor culture in one way is like a regenerating normal culture. If a normal culture with mitoses only around the periphery has a sector cut away, the cells near the V made by removal of the sector begin to divide. It is not the act of cutting *per se* which stimulates cell division. It is the removal of the sector of cells which allows the cells of the new periphery to divide. If cultures are cut but a piece not removed, the cultures heal and control is reestablished (Fischer and Parker, 1929; Ephrussi, 1930; Fischer, 1935).

The Carrel tumor cultures with their gaps had numerous sites for regeneration. They differed from regenerating normal cultures in that the normal cultures formed complete sheets whereas the tumor culture with some cells always dying always had gaps. Here *in vitro* as Nicholson had noted *in vivo* the condition for regeneration was not removed by regeneration.

We now know, as pointed out throughout this book, that the control of differentiation and attendant cell division is from cell to cell and that inhibitory controls fail when there are gaps between cells. Isolation is the pre-condition for regeneration.

There have also developed during the course of evolution control systems in which the regulation is by circulating substances as in metamorphosis or in the control of liver growth (Bucher, Scott and Aub, 1951; Glinos, 1958; Moolten and Bucher, 1967). Some tumors can arise as the result of hormonal abnormalities (Huggins, 1967), but the great majority appear to result from the failure of intratissue communication.

In Carrel's tumor cultures there were actual physical gaps. This is not necessary for the failure of control. In Harrison's experiments with limb fields (Chapter Three), the control of the center of the field over the periphery could be broken by a change in alignment. The same was true when grafts were not aligned properly in *Tubularia* (Chapter Seven). Non-polarized or poorly polarized loosely-adhering cells of tumors could not be expected to communicate properly with each other.

One of the consequences of cell differentiation as indicated in Chapter Eight is that the cells associate with each other in particular patterns. Apparently there are surface configurations which must be properly aligned with neighboring cell surfaces. Even though partially differentiated kidney tubule cells are jumbled with foreign cells, the kidney cells will wander and not come to rest until they are properly aligned with their own kind. They do not come to rest until they have constituted a tubule. This is when they have two free surfaces, a luminal and an external surface, but all lateral surfaces joined with similarly differentiated cells. This is the way a simple tubule is formed. Cells with other surface affinities form other characteristic patterns. In contrast, all tumors show irregularity of patterning. This varies from slight deviation from normal in some of the benign tumors to non-patterning in the anaplastic anarchic tumors (Rupert Willis, 1953).

A failure of cellular affinity has been observed in cultures of cells from anaplastic tumors (Abercrombie and Ambrose, 1958; Abercrombie, 1964). Normal fibroblasts in disperse culture are constantly on the move with a membrane on the leading edge always in motion. When two fibroblasts meet, their membranes stop moving. Other portions of their surfaces may become active and they may glide away from each other. When all sides of a fibroblast have come into contact with other fibroblasts, surface motion stops. In this way large groups of mutually quieting cells can aggregate and tissues form.

The more anaplastic tumor cells behave very differently in culture. Their moving surface membranes do not come to rest when they meet another cell. Instead, the membranes remain restless and the cells do not pause as they meet. One glides over the other. This failure of contact inhibition could account for the property of invasiveness of tumor cells and for the fact that they do not form proper tissues over which control could spread. As a consequence, growth occurs not in the normal pattern of the parent tissue but from multiple loci.

Other evidence for failure of communication between tumor cells has been presented by Werner Lowenstein and Kanno (1967), and Lowenstein and Penn (1967). If an electrical pulse is given to one of two adjacent liver cells, a change in potential can be read off from the other. By the same test, hepatoma cells fail to communicate with each other. This finding may be important because if it is generally true that the agents of control are charged particles moving from cell to cell (Chapters Six and Seven), this failure of sufficient electrical contact between tumor cells may be the basis for the failure of control.

Now after many man-years of work it appears that the fundamental neoplastic change is in a stretch of nucleic acid which is passed on to daughter cells and results in a modified cell surface such that the cells do not aggregate properly and do not make the close contacts over which sufficient repressive information can pass. The result is never-ending multipolar regeneration.

The worst cases are those in which tumor cells are completely dis-aggregated, as in the ascitic tumors. Even in the Ehrlich ascites tumor with dispersed cells there is evidence that a self-limiting control system is at work. The control, however, does not stop the increase in cell number until roughly a billion cells have formed. This number causes the mouse host to die in a few weeks. Robert Burns (1969) has stated the case clearly for the belief that tumors arise as the result of partial failure of the self-repression system. In the past it was guessed that when a tumor began to slow down it was because it was not being supplied enough nutrients by the normal tissues of the host. Now there is considerable evidence (reviewed by Burns) which indicates this is not true. First of all, T. S. Hauschka and coworkers (1957) showed that tumor-cell populations reached the same great number in obese and in starved hosts. Apparently, as in normal tissue, when a certain size is reached cell division decreases to a level which no more than maintains the tumor. It is now apparent that it is not for lack of materials that this happens. Practically all of the cells in an Ehrlich ascites tumor continue to make RNA and protein at a high rate even after the tumor population has stopped increasing. Tumors when compared to normal tissues operate efficiently in extracting raw products such as glucose and amino acids. These supplies seem not to reach too low a level for growth when a large tumor is present.

Most of the tumor cells do stop making DNA, but this is not because the host cannot supply the raw materials. Burns (1968) has demonstrated that a mouse which has produced a billion Ehrlich cells in its abdominal cavity can make four times as many. Burns withdrew by pipette most of the tumor cells from mice with maximal tumors. These cells were making RNA and protein but very few were making DNA. Within a day most of the tumor cells which had remained in the mice were again starting to make DNA. When Burns allowed the tumors to grow several times after removal of most of the cells he learned that a mouse can produce a total of 4 billion tumor cells. Just as in normal tissues, it is the removal of some of the mass which causes regeneration to begin. The difference is that communication between tumor cells is inefficient.

There is another indication that tumors which reach a growth plateau stop growing as the result of self-inhibition rather than as a result of nutrient deficiency. Richard Brown has followed the growth of two different ascitic tumors in the same mouse. The Ehrlich ascitic tumor produces a billion cells when growing alone. The Crocker 180 sarcoma stops growing when a population of two billion cells has been reached. If these tumors were stopping because of a depleted supply of some nutrient, they should not produce as many cells when growing in the same animal where they would be competing for nutrients. Actually, Brown (1970) learned that the Ehrlich plateaus at one billion cells and the Crocker sarcoma at two billion cells even when they are growing together in the same abdominal cavity. This is very good evidence that the growth of each is stopped by its own products rather than by a depletion of nutrients. There is no competition between the two. The total number of cells is the same as the number achieved when the two tumors grow alone.

If growth in these tumors is stopped by their own products, a cure might be devised. The difficulty in most studies of growth control by products acting in a negative feedback loop is that the products are unstable. For example, with the Ehrlich tumor we know from the work of Burns that removal of a large part of the tumor is followed by production of new DNA by the remaining cells within a fraction of a day. The hypothetical repressors at a high concentration at the time of removal of a large part of a tumor are lost rapidly and allow new growth within a day. Sensitive feedback systems can only function if the agent of control is destroyed rapidly. Theoretically, for therapeutic control it would be necessary to have a continuous supply of specific repressors from a large culture of the identical tumor, or a way to make the repressors more stable would have to be devised.

Two very recent studies still not published as this book goes to press give hope for control of tumors by self-repression. The first is a report by P. Bichel that fluid from an abdominal cavity in which a tumor had grown represses the growth of an identical tumor. Robert Burns (personal communication) has repeated this work and finds that he sometimes gets a 50 percent inhibition of tumor growth when subjecting the tumor to its products in abdominal fluid but that the results are not consistent. It appears that the stability of the repressors may vary. It might be wise to try to learn how the algae stabilize tadpole growth repressors (Chapter Seven, Akin, 1966).

Another paper to appear in a short time reports the work of Man-lim Yu and Alfred Perlmutter (1970). They have collected and apparently stabilized growth repressors of two different kinds of tropical fish. The action is quite specific. The repressors from one species, the zebrafish, act to inhibit growth of young zebrafish without slowing down gouramis. In like manner, repressors produced by gouramis act on young gouramis but do not interfere with the growth of zebrafish. The important thing in this work is that the usually evanescent repressors have been stabilized. They were adsorbed by activated charcoal, then taken up in chloroform, evaporated to dryness, and finally dissolved in water. The hope now is that the specific tumor repressors whose action has been demonstrated can be stabilized and used therapeutically.

The discussion above has been about very rapidly growing tumors whose cells do not adhere to each other. At the other end of the spectrum of tumor types are benign tumors whose pattern of aggregation is almost normal. In between lie tumors with all degrees of improper pattern. Rupert Willis (1953), after studying many tumors, generalized that there is an inverse relationship between organized structure and malignancy. In many tumors there is enough resemblance between the neoplastic and the normal structure so that a pathologist can be sure of the organ from which it arose even though it had been found as a metastasis in a far part of the body. In these tumors one can see how the form deviates from normal. For example, a tubular structure can become a multilobed cauliflower-like mass as centers of growth arise too close to other centers. In the case of the crown gall tumors of some plants, it is obvious that dominance has failed and that the tumor is composed of many special meristems with recognizable but abnormal differentiation of leaves around them (Armin Braun, 1952).

The failure of normally patterned transport of repressive information of the sort operating during normal regeneration and maintenance seems to result in never-ending multicentered regeneration.

References Cited

Abercrombie, M. 1964. Cell contacts in morphogenesis. *Arch. Biol.* (Liège) 75: 351–367.

Abercrombie, M. and Ambrose, E. J. 1958. Interference microscope studies of cell contacts in tissue culture. *Exptl. Cell Res.* 15: 332–345.

Bichel, P. 1970. Tumor growth inhibiting effect of JB-1 ascitic fluid. (In press)

Bittner, J. J. 1936. Some possible effects of nursing on the mammary gland tumor incidence in mice. *Science.* 84: 162.

————. 1948. Some enigmas associated with the genesis of mammary cancer in mice. *Cancer Res.* 8: 625–639.

Blum, Harold F. 1948. Sunlight as a causal factor in cancer of the skin of man. *J. Natl. Cancer Inst.* 9: 247–258.

————. 1959. *Carcinogenesis by Ultraviolet Light. An essay in quantitative biology.* Princeton, N.J.: Princeton University Press.

Borek, Carmia and Sachs, L. 1966. *In vitro* cell transformation by x-irradiation. *Nature* 210: 276–278.

Braun, Armin C. 1952. The crown gall disease. *Ann. N.Y. Acad. Sci.* 54: 1153–1161.

Brown, H. R. 1970. The growth of Ehrlich carcinoma and Crocker sarcoma ascitic cells in the same host. *Anat. Rec.* 166: 283.

Bucher, N. L. R., Scott, J. F., and Aub, J. C. 1951. Regeneration of the liver in parabiotic rats. *Cancer Res.* 11: 457–465.

Bünning, E. 1952. Morphogenesis in plants. In *Survey Biol. Progr.* 2: 105–140.

Burmester, B. R. 1952. Studies on fowl lymphomatosis. *Ann. N.Y. Acad. Sci.* 54: 992–1003.

Burns, E. R. 1968. Initiation of DNA synthesis in Ehrlich ascites tumor cells in their plateau phase of growth. *Cancer Res.* 28: 1191–1196.

Burns, E. R. 1969. On the failure of self-inhibition of growth in tumors. *Growth* 33: 25–45.

Carrel, A. 1925. Mechanism of the formation and growth of malignant tumors. *Ann. Surg.* 82: 1–13.

Clayson, D. B. 1962. *Chemical Carcinogenesis.* 468 pp. Boston: Little, Brown.

Coman, D. R. 1944. Decreased mutual adhesiveness, a property of cells from squamous cell carcinomas. *Cancer Res.* 4: 625–629.

Cornfield, J., Haenszel, W., Hammond, E. C., Lilienfeld, A. M., Shimkin, M. B., and Wynder, E. L. 1959. Smoking and lung cancer. Recent evidence and a discussion of some questions. *J. Natl. Cancer Inst.* 22: 173–203.

Duran-Reynals, F. 1942. The reciprocal infection of ducks and chickens with tumor-inducing viruses. *Cancer Res.* 2; 343–369.

————. 1943. The infection of turkeys and guinea fowls by the Rous sarcoma virus and the accompanying variations of the virus. *Cancer Res.* 3: 569–577.

Ebert, J.D. 1959. The formation of muscle and muscle-like elements in the chorio-allantoic membrane following inoculation of a mixture of cardiac microsomes and Rous sarcoma virus. *J. Exptl. Zool.* 142: 587–621.

Eddy, B. E., Stewart S. E., Young, R., and Mider, G. B. 1958. Neoplasms in hamsters induced by mouse tumor agent passed in tissue culture. *J. Natl. Cancer Inst.* 20: 747–761.

Eddy, B. E., Stewart, S. E., Stanton, M. F., and Marcotte, J. M. 1959. Induction of tumors in rats by tissue culture preparations of SE polyoma virus. *J. Natl. Cancer Inst.* 22: 161–171.

Ellerman, V. and Bang, O. 1908. Experimentelle Leukamie bei Hühnern. *Zbl. f. Bak.* 46: 599.

Ephrussi, B. 1930. *Croissance et Régénération dans les cultures des Tissues*. Paris: Masson et Cie.

Fischer, A. 1935. La physiologie de la cellule cancereuse. *Le Cancer*. 12: 160–169.

Fischer, A. and Parker, R. C. 1929. The occurrence of mitoses in normal and malignant tissues *in vitro*. *Brit. J. Exptl. Path*. 10: 312–321.

Foulds, L. 1934. The growth and spread of six filterable tumors of the fowl, transmitted by grafts. *Eleventh Scientif. Rep. Imp. Cancer Res. Fund*. pp. 1–13.

Fürth, J. and Lorenz, E. 1954. Carcinogenesis by ionizing radiation. In *Radiation Biology*, ed. A. Hollaender. vol. 1 part II, p. 1091. New York: McGraw-Hill.

Glinos, A. D. 1958. The mechanism of liver growth and regeneration in *The Chemical Basis of Development*. ed. W. D. McElroy and B. Glass. pp. 813–842. Baltimore: The Johns Hokpins Press.

Gross, L. 1952. Mouse Leukemia. *Ann. N.Y. Acad. Sci*. 54: 1184–1196.

Hauschka, T. S., Grinnell, S. T., Revesz, L., and Klein, G. 1957. Quantitative studies on the multiplication of neoplastic cells *in vivo*. IV. Influence of doubled chromosome number on growth rate and final population size. *J. Natl. Cancer Inst*. 19: 13–31.

Huggins, C. 1967. Endocrine induced regression of cancers. *Science*. 156: 1050–1054.

Kaliss, N. 1958. Immunological enhancement of tumor homografts in mice. A review. *Cancer Res*. 18: 992–1003.

Klein, G. 1966. Tumor antigens. *Ann. Rev. Microbiol*. 20: 223–252.

Lacassagne, A. 1932. Apparition de cancers de la mammelle chez la souris mâle, soumise à des injections de folliculine. *Comptes rendus* 195: 630–632.

Lappé, M. A. 1968. Evidence for the antigenicity of papillomas induced by 3-methylcholanthrene. Doctoral Dissertation. University of Pennsylvania Library.

Loeb, L. 1919. Further investigation on the origin of tumors in mice. VI. Internal secretion as a factor in the origin of tumors. *J. Med. Res*. 40: 477–496.

———. 1945. Chapters 8 and 9, pp. 89–115, in *The Biological Basis of Individuality*. Springfield, Illinois: Thomas.

Loewenstein, W. R. and Kanno, Y. 1967. Intercellular communication and tissue growth. I Cancerous growth. *J. Cell Biol*. 33: 225–234.

Loewenstein, W. R. and Penn, R. D. 1967. Intercellular communication and tissue growth. II Tissue regeneration. *J. Cell Biol*. 33: 235–242.

Lucké, B. A. 1938. Carcinoma in the leopard frog: its probable causation by a virus. *J. Exptl. Med*. 68: 457–468.

Lwoff, A. 1953. Lysogeny. *Bact. Rev*. 17: 269–337.

McKinnell, R. G. 1969. Lucké renal adenocarcinoma: epidemiological aspects.

in *Biology of Amphibian Tumors*, ed. M. Mizell. Recent Results in Cancer Research, Special Supplement. pp. 254–260. New York: Springer-Verlag.

Mizell, Merle, ed. Tulane Amphibian Tumor Conference. Berlin: Springer-Verlag, 1969.

Möller, G. 1963. Studies on the mechanism of immunological enhancement of tumor homografts. I Specificity of immunological enhancement. *J. Natl. Cancer Inst.* 30: 1153–1175. II Effect of isoantibodies on various tumor cells. *J. Natl. Cancer Inst.* 30: 1177–1203.

Moolten, F. L. and Bucher, N. L. R. 1967. Regeneration of rat liver: transfer of humoral agent by cross circulation. *Science* 158: 272–274.

Moulton, F. R. 1945. Editor of *A Symposium on Mammary Tumors in Mice.* AAAS, Washington. 223 pp.

Mulcare, D. J. 1969. Non-specific transmission of the Lucké tumor. in *Biology of Amphibian Tumors*, ed. M. Mizell. Recent Results in Cancer Research, Special Supplement. pp. 240–253. New York: Springer-Verlag.

Murphy, J. B. 1935. Experimental approach to the cancer problem. *Bull. Johns Hopkins Hosp.* 56: 1–31.

Medical Research Council. 1954. Tobacco smoking and cancer of the lung. *Brit. Med. Jour.* 1: 1523–1524.

Nicholson, G. W. de P. 1950. Chapter 13 in *Studies on Tumour Formation.* pp. 307–333. London: Butterworth.

Pott, P. 1779. *Chirurgical Works* 3: 225.

Prehn, R. T. 1965. Role of immunity in the biology of cancer. *Fifth Nat. Cancer Conf. Proc.* pp. 97–104. Philadelphia: Lippincott.

Rafferty, K. A., Jr. 1964. Kidney tumors of the leopard frog. A review. *Cancer Res.* 24: 169–185.

Rose, S. M. 1952. A hierarchy of self-limiting reactions as the basis of cellular differentiation and growth control. *Amer. Nat.* 86: 337–354.

Rose, S. M. and Rose, F. C. 1952. Tumor agent transformations in amphibia. *Cancer Res.* 12: 1–12.

Rous, P. and Murphy, J. B. 1914. On the causation by filterable agents of 3 distinct chicken tumors. *J. Exptl. Med.* 19: 52–69.

Sachs, L. 1965. A theory on the mechanism of carcinogenesis by small deoxyribonucleic acid tumour viruses. *Nature* 207: 1272–1274.

Study Group on Smoking and Health. 1957. Joint report. *Science* 125: 1129–1133.

Snell, G. D. 1954. The enhancing effect (or actively acquired tolerance) and the histocompatibility—2 locus in the mouse. *J. Natl. Cancer Inst.* 15: 665–675.

Strong, L. C. 1935. The establishment of the C3H inbred strain of mice for the study of spontaneous carcinoma of the mammary gland. *Genetics.* 20: 586–591.

Tweedell, K. S. 1955. Adaptation of an amphibian renal carcinoma in kindred races. *Cancer Res.* 15: 410–418.

Tweedell, K. S. 1969. Simulated transmission of renal tumors in oocytes and embryos of *Rana pipiens*. In *Biology of Amphibian Tumors*, ed. M. Mizell. Recent Results in Cancer Research, Special Supplement. pp. 229–239. New York: Springer-Verlag.

Willis, R. A. 1953. *Pathology of Tumors* (2nd edit.) London: Butterworth & Co.

Yamagiwa, K. and Koichi, I. 1918. Experimental study of the pathogenesis of carcinoma, *Am. Jour. Cancer*. 3: 1–29.

Yu, Man-lim and Perlmutter, A. 1970. Growth inhibiting factors in the zebrafish, *Brachydanio rerio* and the blue gourami, *Trichogaster trichopterus. Growth* 34:153–175.

Index

251